Microsystems for Bioelectronics

Microsystems for Bioelectronics:
The Nanomorphic Cell

Victor V. Zhirnov

Ralph K. Cavin III

AMSTERDAM • BOSTON • HEIDELBERG • LONDON • NEW YORK • OXFORD
PARIS • SAN DIEGO • SAN FRANCISCO • SINGAPORE • SYDNEY • TOKYO
William Andrew is an imprint of Elsevier

William Andrew is an imprint of Elsevier
The Boulevard, Langford Lane, Kidlington, Oxford OX5 1GB, UK
30 Corporate Drive, Suite 400, Burlington, MA 01803, USA

First edition 2011

British Library Cataloguing in Publication Data
A catalogue record for this book is available from the British Library

Library of Congress Cataloging-in-Publication Data
A catalog record for this book is availabe from the Library of Congress

ISBN–13: 978-1-4377-7840-3

For information on all William Andrew publications visit our web site at books.elsevier.com

Printed and bound in the USA

11 12 13 14 15   10 9 8 7 6 5 4 3 2 1

Contents

Preface

MOTIVATION

This book is focused on questions that arise in the design of functional electronic systems that have dimensions of a few microns. Is it physically reasonable to entertain the idea that such systems might ultimately exist? The designation, *nanomorphic cell*, has been used to convey the thought that the functionality and size of such a system is defined by what the system must accomplish and by the environment within which it must operate. The prefix *nano* is used to suggest that components of the micron-scale system will likely have nanometer dimensions. The word *morphic* literally means *in the shape of* and its use is to convey the idea that the intended application shapes the functionality of the nanomorphic cell. Typically, the system volume must accommodate an energy provider, sensors and communication, and computation and memory – all within just a few cubic microns. To help fix ideas, it is imagined that an application for the nanomorphic cell might be to collect in vivo information on the health of living cells that it encounters and to communicate its findings to an external decision-maker. Of course, the application space for the nanomorphic cell would be much larger than this illustration.

What gives rise to the hope that a nanomorphic system might be possible? One source of hope is that the scaling of integrated semiconductor components will almost certainly continue. Leading-edge manufacturing as this book is written produces devices whose minimum feature size is on the order of 25 nanometers and there is the expectation that devices with minimum feature sizes of five nanometers may be feasible. The number of devices on a one centimeter by one centimeter chip exceeds one billion and will certainly approach several hundred billion when the physical scaling limits for CMOS technology are reached. A second reason to believe that nanomorphic systems may be feasible is the trend to produce *systems-on-a-chip* whose components are not restricted to electrical logic gates and memory systems but also can include other components such as sensors, communication components, and energy sources. Although current systems-on-a-chip are far larger physically than the proposed nanomorphic system, the integration of elements that function in different physical domains is already taking place.

As will become evident in the book, available energy defines what a nanomorphic cell could accomplish and thus it is reasonable to ask whether there are existing systems that could serve as inspiration for the nanomorphic cell. One such example is the living cell whose physical size is comparable to the nanomorphic cell and whose complexity and energy efficiency for its operations are unparalleled. It is fair to say that the objectives of the living cell differ somewhat from those of the nanomorphic cell, but the analogy of the living cell as a special kind of computer is not too far-fetched.

PHILOSOPHY

The nanomorphic cell is inescapably multidisciplinary and for the authors the writing challenge has been to discuss succinctly a broad range of disciplines in such a way that physical insights into performance are manifest. The goal was to write the book in such a way that it is useful to a broad audience of workers in biology, chemistry, physics, medicine, and engineering. The vocabulary across all of these fields is necessarily vast and it is therefore important to carefully develop terminology that supports the exchange of ideas across these diverse disciplines. Throughout, an effort has been made to utilize basic physics, chemistry, biology, materials, and engineering concepts in order to make the book more broadly useful. At the same time, the models that have been introduced are sufficiently powerful so that insights into the limits of particular technologies should be evident. Often, these approximate models can be refined by the inclusion of appropriate pre-factors due to their fidelity to the science upon which the model is based. The authors have successfully utilized sections from the book in a multidisciplinary graduate class in engineering and the students from several disciplines have had no difficulty grasping the concepts offered by the book. The book will best serve the reader with an introductory background in physics, chemistry, and biology.

The reader will notice that the style of writing in the book is a mix between the expository style of textbooks and the tightly referenced style of scientific articles. Wherever possible, proofs are offered based on fundamental science in an effort to provide the reader with insights into the physical limits for the various technologies that are discussed. However, the breadth of the topics covered requires that contact be made with the scientific literature in each of the topical areas to provide quantitative data and to give a sense of the status of the various technologies that are discussed. An effort has been made to briefly present these literature references in an orderly manner that relates to and supports the expository material.

OBJECTIVES

Each of the chapters develops conclusions using 'best-case scenarios' that are constructed so as to be consistent with the fundamental laws of physical sciences. The fundamental limits that result, for say energy sources, often stimulate a response from the audience that is something like "Have you considered ...?" This is exciting since it indicates that, upon hearing the claimed limitation, the listeners are stimulated to suggest alternative scenarios. Indeed, the reader is encouraged to read each chapter critically since the results given represent the current understanding of fundamental physical limits and new knowledge about the physical world is constantly resulting from research. The primary objective of this book is to stimulate the reader's creative response to the challenges that are proposed and innovative solutions are always welcome.

STRUCTURE

Chapter 1 introduces the concept of the nanomorphic cell, and makes a reference to the state-of-the-art of the integrated microsystems. Chapter 2 is devoted to a study of the potential of various energy sources whose dimensions are in the 10-micron range. The volume constraint for the nanomorphic system forces the search for extremely scaled energy sources and energy conversion mechanisms. Chapter 3 considers the projected capability of computation and memory systems when ultimately scaled devices are used. It is shown that energy barriers are a fundamental concept in defining the ultimate scalability of transistors and a discussion of the physics of device-to-device communication is also included. In Chapter 4, an overview of biological sensors, again limited to 10 microns in size, is undertaken with an emphasis on sensitivity, selectivity, and settling times for various sensors. Emphasis is on electrical, biochemical, and thermal sensors operating in an environment of thermally induced noise. Chapter 5 discusses the limits of electromagnetic communication systems when the systems are constrained to dimensions on the order of 10 microns. It turns out that the communication of a bit of information over distances of the order of 1 meter is much more costly from an energy expenditure perspective than any other activity of the nanomorphic cell. Chapter 6 focuses on the comparison of the computational capabilities of the in silico nanomorphic cell with those of the in carbo living cell. Chapter 6 is very multidisciplinary since it requires the integration of data from many sources to enable a model for the living cell as a computer. Indeed, in carbo systems are the most dramatic existence proof for functional micron-scale systems.

Throughout the text, an effort is made to rely primarily on basic physics and chemistry to provide insight into the limits that are set forth. The goal is to provide fundamental understanding of the underlying mechanisms that limit the nanomorphic cell capabilities. The authors agree that more detailed models would provide more refined estimates for the various limits but believe that if at all possible, a physical understanding of limiting factors should precede detailed calculations.

NOTATION

Due to the multidisciplinary nature of the book, notation from several different scientific areas was required. It sometimes happens that different fields will utilize different notation for the same physical variable. The decision was made to utilize notation consistent with that commonly used in the particular discipline in each chapter to the greatest extent possible. Therefore, a list of symbols with definitions is provided for each chapter.

Acknowledgment

The authors gratefully acknowledge the support and encouragement for this project provided by our colleagues at the Semiconductor Research Corporation. Ms. Valerie Eng and Ms. Anastasia Batrachenko read critically each of the chapters, and their editorial comments and criticism helped the authors to better comprehend where improvements were needed. Their detailed reviews greatly improved the book.

CHAPTER

The nanomorphic cell

1

CHAPTER OUTLINE

1.1 INTRODUCTION

Suppose that it is desired to design an autonomous system embedded in the human body whose mission is to analyze the health of cells that it encounters and to report its findings to an external agent. The living cell, which is an organic autonomous system, provides an existence proof that functional and autonomous systems are possible at the scale of a few microns. This text investigates the feasibility of the design of a functional inorganic system on the same physical scale as the living cell, i.e., with overall dimensions of several microns. One reason to believe that such a design might be possible is the remarkable progress that has been made in technologies for semiconductor chips, where some of the devices on the chip already have dimensions on the order of a few nanometers, and dimensional scaling is anticipated to continue for a few more generations. In addition, there is a trend to incorporate more functionality onto a single chip by including devices whose domains of operation are not only electrical but also mechanical, thermal, chemical, etc. These 'system-on-a-chip' designs may point the way to integrated chips with increasing degrees of functionality. The term 'nanomorphic cell' is used herein to reflect the fact that emphasis is on inorganic integrated systems whose inspiration is derived from their biological counterparts. (The term 'morphic' literally means 'in the shape of'.)

To help fix ideas, imagine that the nanomorphic cell is to be injected into the body to interact with the living cells and to support certain diagnostic and/or therapeutic actions. In order to do this, it is stipulated that the nanomorphic cell must acquire data indicative of the health of the living cells that it contacts, analyze the sensed data, and communicate its findings to an external agent. Since the nanomorphic cell is untethered, it must either harvest energy from its surroundings or carry an embedded energy source.

Subjectively, it seems reasonable to postulate that an embedded system of this size would contain only minute and harmless amounts of materials that in larger quantities might be harmful to the body and, furthermore, that the normal body waste disposal processes could manage the removal of nanomorphic cells when they have reached the end of their useful lives. The nanomorphic cell would need to employ some sort of triggering mechanism to signal its elimination from the body. Of course, this is all hypothetical and would need to be verified, e.g., by careful toxicology studies. The in vivo functional nanomorphic cell is used as an example throughout the text as a vehicle to motivate the study of the impact of extreme scaling on system component performance limits.

1.2 ELECTRONIC SCALING

Electronic circuits and systems are constructed from a number of components, the most basic of which is the semiconductor transistor (see Chapter 3), that is used in digital applications as a binary switch. Tremendous progress has been achieved in reducing the physical size of semiconductor transistors – within the last 40 years, the number of transistors in a ~1 cm^2 integrated circuit (IC) chip increased from several thousand in the 1970s to several billion in 2010. The long-term trend of transistor scaling is known as Moore's Law: The number of transistors in an IC chip approximately doubles every two years (see Fig. 1.1a). Moore's Law has been one of the major drivers for the semiconductor industry. The increased complexity of the ICs, accompanied by exponential decline in cost per function, has resulted in increased functionality and expansion of application space of semiconductor products.

The tremendous increase in the number of transistors per chip was enabled by scaling – development of technologies to make smaller and smaller transistors, whose critical feature size decreased from ~10 μm in the 1970s to ~20 nm (transistor gate length, L_g) in 2010 (for brief definitions of feature sizes see Box 1.1). According to the International Technology Roadmap for Semiconductors (ITRS) [1], feature sizes may be as small as ~5 nm by 2022 (the trend of decreasing critical device size is shown in Figure 1.1b). At this *nanoscopic* scale, the properties of matter may be different from those of bulk materials and the physical effects of these *nanomaterials* may play a role in *nanodevice* operation. Fabrication of such tiny structures requires significant technological advances, thus *nanofabrication*. The entire field is often called *nanotechnology* and the subfield related to electronics is called *nanoelectroncs*. To be sure, the chips themselves typically have a footprint on the order of 1 cm^2; in other words the current and most of anticipated nanotechnological solutions are implemented with devices with nanometer features while the integrated system features are in the millimeter range.

While the scaling limits of individual devices can be estimated from physical considerations [2], the question remains open of how small a functional system can be and still offer useful functionality. To address this question, it is important to understand not only the device characteristics, but also how connected systems of these devices might be used to perform complex functions. In the context of this text, a 'system' performs a number of 'functions' (F) in response to external stimuli, which are taken to be information flows (I). A schematic representation of a microelectronic system is shown in Figure 1.2. It contains six essential units:

S – A sensing unit that receives inputs (information) from the outside world.
A – An actuator that performs an 'action' on the outside world.
C – A communication unit that transmits information to the outside world.
M – A memory unit that stores instructions, algorithms, and data.

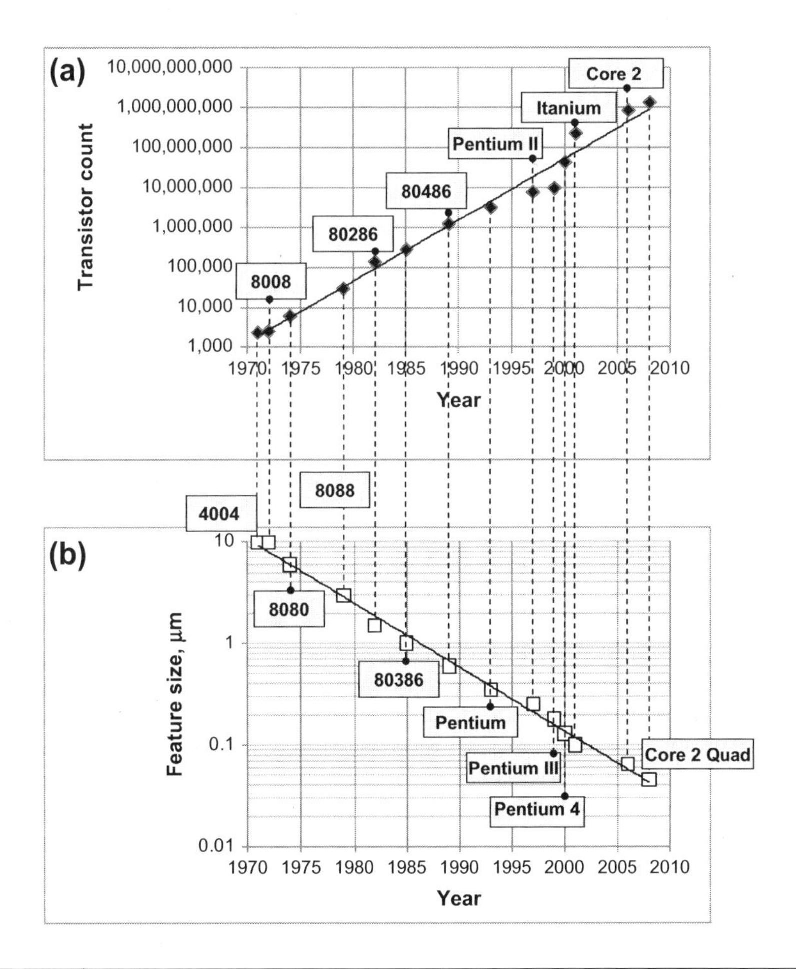

FIGURE 1.1

(a) Transistor counts for Intel microprocessor integrated circuits in 1971–2009; (b) Corresponding critical feature size, *F*, in ICs (see Box 1.1)

BOX 1.1 SEMICONDUCTOR ROADMAP

A very useful resource for estimating the progression of semiconductor IC technology is the International Technology Roadmap for Semiconductor (ITRS) [2], which is updated biannually. Generation of ITRS is a worldwide endeavor and it represents a consensus 15-year technology forecast from a worldwide group of leading industrial technologists – experts from academia and government agencies. The 2007 edition of ITRS forecasts FET minimum feature size (gate length L_g) of ~5 nm and transistor density of ~10^{10} transistors/cm^2 for high-performance microprocessor chips [2]. In addition to gate length, another characteristic feature size is the 'half pitch' *F*, which is equal to the width of the interconnect wires in the memory array (and the separation between the neighbor wires). For example, in 2010 Lg~20 nm and F~45 nm.

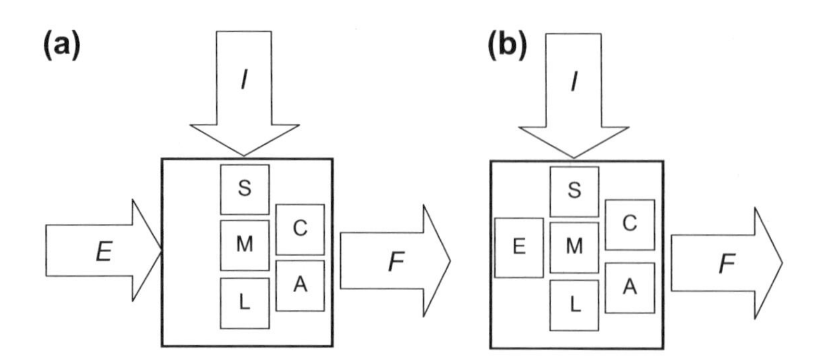

FIGURE 1.2

Block diagrams for arbitrary electronic system: (a) with external energy supply; (b) with internal energy source

L – A logic unit that processes information collected from **A** and that provides summary information for transmission by **C.**

E – An energy source that powers operations of all units. The energy can be supplied to the system from an external energy source (Fig. 1.2a) or/and the system can have an internal energy source (Fig. 1.2b). In the former case, an energy converter must be embedded into the system. In the latter case a finite supply of energy is implied.

Each of the above essential system units occupies certain volume in space and, taken together, they determine the scaling limits of an electronic system. Investigation of these limits is the primary purpose of this text.

1.3 NANOMORPHIC CELL

The *nanomorphic cell* concept refers to an atomic-level, integrated, self-sustaining microsystem with six primary components: energy supply, sensing, actuation, computation, memory, and communication. It is a model system, designed to analyze the physical scaling limits of electronic systems, which for future reference is postulated to be confined within a 10 μm × 10 μm × 10 μm cube [3]. A cartoon for the hypothetical nanomorphic cell is shown in Figure 1.3.

Volume is one of the primary design constraints for the nanomorphic cells and it will be shown in the subsequent chapters that this resource should be very carefully allocated among all functional units. Literally every atom must play a role when one needs to fit all functional units into the 10 μm × 10 μm × 10 μm cube (Fig. 1.4). From an application point of view, the nanomorphic cell can be considered as an extreme example of a class of systems known generically as autonomous microsystems, for example WIMS (wireless integrated microsystems) [4], PicoNode [5], Lab-on-a-Pill [6], and Smart Dust [7]. A brief discussion of these will be given at the end of the chapter.

1.3.1 The nanomorphic cell vis-à-vis the living cell

The living cell is a marvelous machine that is the cornerstone of all living things and that has embedded mechanisms for (i) staying alive and for (ii) reproducing itself. It is not an exaggeration to

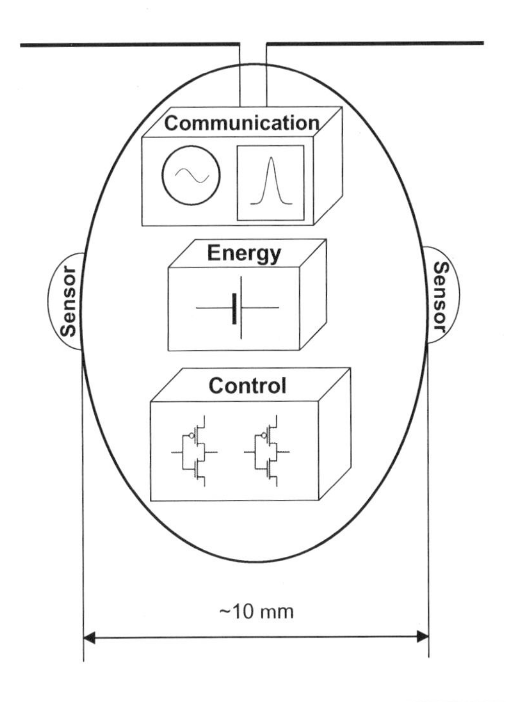

FIGURE 1.3

Cartoon for the *nanomorphic cell* showing essential components and physical scale

say that the living cell is purposeful [8]. In order to achieve the goal of staying alive, cells need to acquire information about their environment. Next, the cell needs to appropriately respond to the information, e.g., about changes in temperature, water availability, nutrients supply, dangerous species, and many other factors. The ability to acquire and use such information is critical for organism survival [9]. As was recently stated by A. Kinkhabwale and P. Bastiaens (Max Planck Institute of Molecular Physiology), 'to live is to compute' [8].

The second goal of reproduction itself involves a series of precise tasks which requires controlled flows of information and matter ensuring that all atoms forming the cell are positioned in a specific place within the cell. Recently, Antoine Danchine at the Institut Pasteur in Paris, France suggested that a living cell (e.g. a bacterium) can be considered *a computer making computers* [10]. The author argues that a cell has all essential attributes of a computer, i.e. a machine expressing a program.

The studies of the living cell as a functional microsystem may help engineers to understand the physical limits of scaling for functional electronic systems. Or, vice versa, lessons from extremely scaled electronic systems may help biologists to gain new insights into the fundamental questions of theoretical biology, such as 'What is the minimum size of Life?' [11], or 'What is the minimum energy needed to support Life?' [12].

In this section, a brief overview of unicellular organisms is presented with an emphasis on system size and energy. Other system-level parameters such as 'speed of operation', informational complexity, etc. will be discussed in Chapter 6.

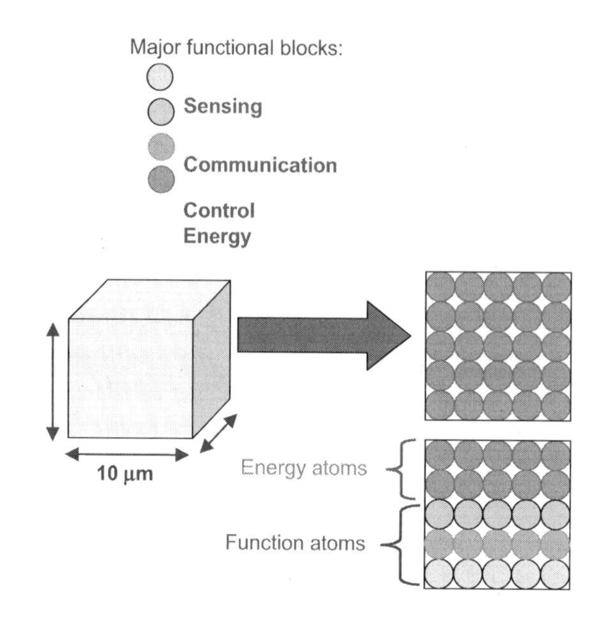

FIGURE 1.4

An atomistic view on *nanomorphic cell*: Very limited space needs to be shared by sensors, power supply, and electronic components. At this scale, every atom must play a role

Functional blocks in the living cell are complex biomolecules, such as proteins, DNA, and RNA (see Box 1.2).

The two main groups of living cells, the *prokaryotes* and *eukaryotes*, are shown in Figure 1.5. The *prokaryotes* (Fig. 1.5a,b) and *eukaryotes* (Fig. 1.5c) are commonly distinguished by the presence (eukaryotes) or absence (prokaryotes) of an internal membrane, encapsulating the DNA, thus forming a *nucleus* inside the cell. Another key difference is that eukaryotes utilize a number of specialized subunits (organelles) in their operations. One important subunit is the *mitochondrion*, the internal cell's power supply. It is sometimes argued that the presence of mitochondria is the main distinction between prokaryotes and eukaryotes [15].

BOX 1.2 MACROMOLECULAR ELEMENTS IN THE CELL

Proteins – Macromolecules with typical dimensions of a few nanometers that serve as basic building blocks of living organisms. There are thousands of different types of protein in a cell. Different proteins play different roles in the cell. Structural proteins provide 'bricks' to form the living tissues. Enzymes are proteins that catalyze biochemical reactions. Many proteins in living cells appear to have as their primary function the transfer and processing of information [13, 14], including sensing, logic, and short-term memory functions.

DNA (deoxyribonucleic acid) – A macromolecule whose primary function is the long-term storage of information about the cell. This information controls cellular metabolism, replication, and overall survival functions.

RNA (ribonucleic acid) – A type of macromolecule whose primary roles in the cell are to read-out information encoded by DNA, to carry this information to certain locations in the cell, and to facilitate protein synthesis. There are several different kinds of RNA specialized for different functions.

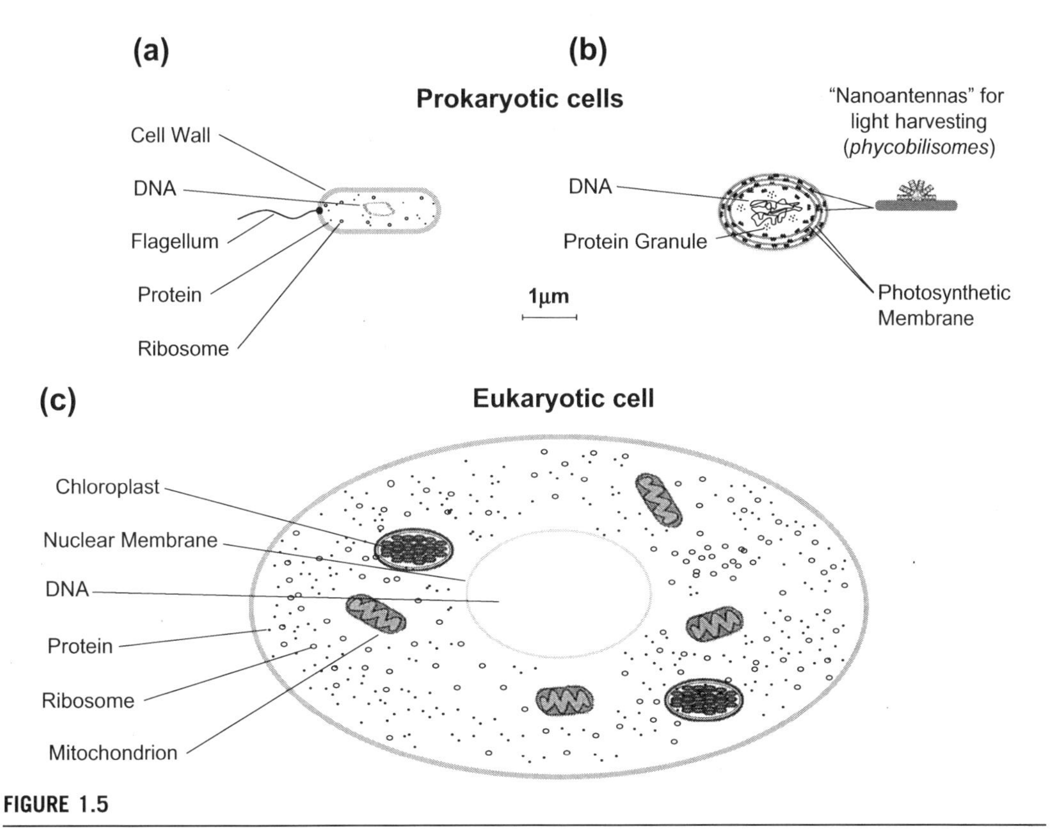

FIGURE 1.5

Two primary groups of living cells (shown at approximate relative sizes): (a, b) *prokaryotic* cells ((a) bacterium, (b) cyanobacterium) and (c) *a eukaryotic* cell

While the cells are very complex structures consisting of many components, several of the most essential components can be identified in analogy to the autonomous microsystem parts shown in Figure 1.2. All cells have a cell envelope, which protects structural integrity of the cell. Inside of the envelope, there is a *memory unit* consisting of DNA molecules. This unit stores all information about the cell, which is used, e.g., for the cell's replication. The *logic unit* of the cell consists of two major groups of 'molecular devices': (1) different types of RNA molecules, which read-out information from the DNA and carry this information to certain locations in the cell and (2) proteins, many of which have as their primary function the transfer and processing of information, including sensing, logic, and short-term memory functions. The cell also contains *ribosomes*, which also can be regarded as part of the cell's *logic unit* as will be discussed in Chapter 6. A ribosome is a biomolecular formation that consists of proteins (~35%) and RNA (~65%). The size of ribosomes is ~20 nm. The function of ribosomes is to synthesize the protein molecules according to commands from DNA (transferred by the messenger RNA).

Typical prokaryotes are *bacteria* (see Box 1.3). The bacteria do not have significant internal energy sources and they almost entirely rely upon the energy from their environment, e.g. dissolved nutrients. It is commonly argued that smaller cells are more efficient at resource uptake due to the greater

BOX 1.3 UNICELLULAR ORGANISMS

Bacteria and Archea – The prokaryotic single-cell organisms, the smallest among all living species. The total number of prokaryotes on Earth is estimated to be 4–6 × 10^{30} cells. The corresponding amount of their cellular carbon is ~3.5–5.5 × 10^{14} kg, which constitute 60–100% of the estimated total carbon in plants [16]. Probably the most studied bacterium is *Escherichia coli* (*E. coli*) with size of ~1 μm. It is often considered the model organism in microbiology.

Cyanobacteria – The bacteria that utilizes photosynthesis to power itself. For example, *Prochlorococcus marinus* (*P. marinus*), which is among the smallest photosynthetic organisms known today (cell size 0.5–0.7 μm), is also the dominant photosynthetic organism in the ocean [25].

Eukariotic microalgae – Smallest single-cell plants. A typical example is *chlorella*. Microalgae cells contain photosynthetic organelles, the chloroplasts, and they play a very important role for the life on Earth, as they produce ~50% of the atmospheric oxygen.

Protozoa – Single-cell animals. A typical example is *ameba* (amoeba). Another interesting single-celled organism is *euglena* that exhibits both plant and animal traits, and can obtain energy from sunlight through photosynthesis, or consume nutrients from organic materials in its exterior.

surface-to-volume ratio – this is one of the reasons the prokaryotes, e.g. bacteria, are among the smallest living species with a typical size of ~ 1 μm. Also it makes motility an essential attribute of bacterial microsystem, as a permanent search for energy is mandatory. Bacteria commonly have a locomotion actuator, the *flagellum* (Fig. 1.5a).

Another type of prokaryotic cell is *cyanobacteria*, whose distinct feature is an internal energy source based on light harvesting (photosynthesis). The photosynthetic 'unit' consists of a system of internal concentric membranes (*thylakoids*) containing photosynthetic pigments, such as *chlorophylls*. Special light-harvesting antennas (*phycobilisomes*) are attached to the photosynthetic membranes, which enhances the efficiency of the light-harvesting process. Cyanobacteria do not have flagella, apparently due to their internal energy source which makes the motility function not as critical as in bacteria.

Eukaryotic cells, typically about 10 μm in size, are larger than prokaryotes. Their larger size allows them, in principle, to draw on multiple energy sources, such as chemical and solar energy. Specifically, eukaryotes have special organelles acting as internal power supplies such as *mitochondria* for chemical energy conversion and *chloroplasts*, which utilizes solar energy through photosynthesis. An essential feature of both mitochondria and chloroplasts is that they possess their own separate DNA genomes that control the electron transport in energy-producing reactions. In other words, both mitochondria and chloroplasts act as precisely controlled 'smart' energy sources for biological cells. Also eukaryotic cells can have significant internal energy storage typically in the form of glucose polymers such as of starch (plant cells) or glycogen (animal cells). Multicellular organisms (including the human body) consist of a large number of eukaryotic cells. Note that the two main categories of the living cells shown in Figures 1.5a,b and 1.5c, if viewed as autonomous microsystems, can be related to the two classes of the microsystems shown in Figures 1.2a and 1.2b. Box 1.3 contains brief definitions for some of the common classes of single-cell organisms.

1.3.2 Cell parameters: Mass, size, and energy

The cell consumes energy in order to interact with its environment and build another cell, i.e. replicate. In addition, according to the second law of thermodynamics, all complex non-equilibrium systems are

Table 1.1 Examples of mass, size, and power consumption of the cell, averaged within different taxonomic groups

Cell average within different taxonomic groups	Wet mass, kg	Volume, μm^3	Size, μm	Power, W
Prokaryote	2×10^{-15}	2	1.3	10^{-14}
Cyanobacteria	7×10^{-14}	70	4	2.5×10^{-13}
Eukariotic microalgae	6×10^{-12}	6000	18	5×10^{-11}
Protozoa	3×10^{-11}	3×10^4	32	2.5×10^{-10}
Human cells	10^{-12}	1000	~10	4×10^{-12}

(Source: [12])

subject to spontaneous degradation. Therefore living organisms must spend some energy to fight accumulating disorder, for example to repair damage by replacing some 'broken' macromolecules, etc. [17].

A characteristic measure of the rate of energy consumption, or power, by living organisms is their metabolic heat rate. Studies across a broad spectrum of the organisms have revealed a striking observation that their mean metabolic rates are confined in a narrow range between 0.3 and 9 Watts per kilogram of body mass [12]. Table 1.1 presents examples of mass, size, and power consumption of the cell, averaged within different taxonomic groups (the numbers were derived from [12]).

It is interesting to estimate the rate of energy consumption of the cellular structure of the human body. First, note that approximately 70% of the living cell ('wet mass') is due to water. Therefore as an order-of-magnitude estimate, it is reasonable to assume that the cell has a 'wet' mass density close to the density of water (1 gram/cm^3). Thus, a cell with a size of 10 μm has a mass of 1 ng or 10^{-12} kg. Now, since a typical weight of a human can range between 50 and 100 kg, an approximate number of cells in the body is 5×10^{13}–10^{14}, as is consistent with numbers in the literature. Next, since the average power consumption per cell in mammals is 4×10^{-12} W (Table 1.1), multiplying this rate by the number of cells results in 200–400 W for human power consumption. By comparison, the average power per person can be derived from medical studies of daily energy intake as follows. The typical total energy expenditure of men aged 30–39 years is 14.3 MJ/day = 3400 kcal/day [18]. Thus the corresponding average power is $P = E/t = 14.3$ MJ/(24 h × 3600 s) ~165 W.s.

The two different estimates for human power consumption differ by about a factor of 2. The difference is in part due to variation in the size and in energy consumption rates of various human cells. Also, the number assumed for human cell power in Table 1.1 is an average across many different mammals [12].

1.4 CURRENT STATUS OF TECHNOLOGIES FOR AUTONOMOUS MICROSYSTEMS

1.4.1 Cardiac pacemakers

This class of devices represents an example of successful commercialization for implantable bioelectronic systems. It also is a clear illustration of how the success of miniaturization in semiconductors enabled new application domains. In fact, the development of an implantable cardiac pacemaker entirely depended on the commercialization of the transistor in the 1950s allowing the design of miniature battery-powered electrical circuits [19]. Thanks to Moore's law, the complexity of the semiconductor components of pacemakers has tremendously increased with time, while their size and mass have decreased (Fig. 1.6). Modern pacemakers contain microprocessors, memory, different sensors, data transmission, etc. It should be noted, however, that the further scaling of the pacemaker system size has decreased in recent years mainly due to the limited scalability of power supplies (modern pacemakers weigh ~20 g and have a volume ~10 cm^3).

1.4.2 Lab-on-a-pill: Wireless capsule endoscopy systems

Currently, there are several commercially available wireless capsule endoscopy systems [21, 22]. The endoscopes typically contain miniature video cameras, an LED light source, a radio transmitter (with antenna), and batteries [21]. A typical size of the 'pill' envelope is ~1 × 3 cm, with an approximate volume of ~ 10 cm^3 and weight of ~4 g [21].

The swallowable capsules can also contain sensors, e.g. pH, temperature, and pressure [21]. A representative example is a miniaturized device enabling real time in situ measurements in the gastrointestinal (GI) tract that was reported in [6]. It contains temperature (silicon diode) and pH (ISFET) sensors along with an integrated radio transmitter using 433.92 MHz frequency. The radio transmitter is equipped with an integrated loop antenna (1/16 λ) that sends data to a base station connected to a computer. The device also contains a permanent magnet used for non-visual tracking of the system. The system control unit was fabricated on a 44-pin 4.5 mm × 4.5 mm silicon die using 0.6 μm CMOS process. The encapsulated lab-on-a-pill was 1.2 mm × 3.6 cm in size, weighed 8 g, and was powered by a 3 V power supply consisting of two Ag_2O battery cells SR48 (75 mAh). During operation, the data transmission distance was limited to ~1 m. The power consumption of the device was 4.75 mW in the standby mode and 15.5 mW in the data transmission mode.

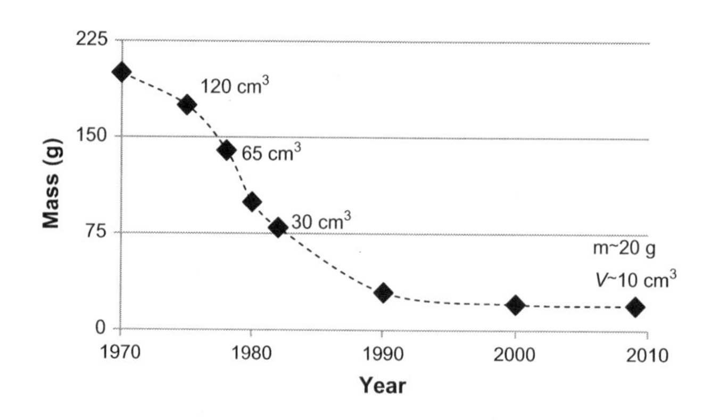

FIGURE 1.6

Pacemaker mass/size trends (adopted and updated from [20])

1.4.3 'Smart Dust'

The term 'Smart Dust' is used to describe different wireless sensor network devices of ~1 mm^3 in size [7, 23]. An individual wireless sensor node typically contains sensors, hardware for computation and communication, and a power supply [7]. These nodes (also called 'motes') are assumed to be autonomous and programmable and are able to communicate with each other. Originally, the concept was driven by envisioned military applications. It was recently proposed that a network of autonomous sensor nodes could form a ubiquitous, embedded computing platform [5, 7]. Operational prototypes of such modes on the order of ~ 1 cm^3 have been demonstrated. One of the smallest motes using passive optical communication was only 16 mm^3 in volume and was powered by a solar-cell source [26]. The challenging target of demonstrating a fully operational mote 1 mm^3 in size has yet to be achieved. The most difficult components to scale are energy sources and RF communication components.

1.4.4 Wireless integrated microsystems (WIMS)

In WIMS, the emphasis is on adding non-electrical components to the wireless sensor microsystems [4, 24]. The combination of semiconductor and microelectromechanical elements (MEMS) into one hybrid system considerably enhances system functionality and expands the application space. An important application space is in the area of neural probes, where electronics is integrated with micrometer-scale multi-electrode arrays [24]. Another WIMS example is miniature high-resolution pressure sensors that are used, for example, for cardiovascular measurements. Perhaps one of the most impressive examples of a hybrid WIMS system is a prototype gas chromatography system, for high-sensitive chemical detection. The demonstrated prototype of this complex automated system had a volume of only ~200 cm^3 (on the order of $10 \times 10 \times 2$ cm) and further developments are intended to eventually achieve wristwatch implementations [4]. In general, the combination of embedded signal processing and wireless communication with MEMS is enabling broad new applications, covering different areas, including homeland security, health care, food and environmental monitoring, etc.

1.5 CONCLUDING REMARKS

Scaling of semiconductor device technology provides encouraging opportunities for making new functional systems across a broad application spectrum. Electronic devices, such as transistors, can be made very small, ultimately exhibiting feature sizes of a few nanometers. However, while corresponding miniaturization efforts for systems are currently underway, the overall system sizes remain relatively large, typically on the centimeter scale. As a general rule, functional systems are more difficult to miniaturize than devices. A practical shorter-term goal for a small autonomous microsystem would be to achieve ~1 mm^3 volume. Moreover, while the scaling limits of individual electronic devices have been estimated from physics-based considerations, the question of how small a functional system can be and still offer useful functionality remains open.

The focus of this text is on autonomous systems whose dimensions are two orders of magnitude smaller than today's 1 mm target. Thus, a focus of the ensuing chapters is on physical limits for the various components of the nanomorphic cell. The *nanomorphic cell* abstraction proposed in this book may allow one to analyze the fundamental limits of attainable performance for nanoscale systems in much the same way that the *Turing machine* and the *Carnot engine* support such limit studies for information processing and heat engines respectively.

Since the living cell, which is an organic autonomous system, provides an existence proof that functional and autonomous systems are possible at the scale of a few microns, the size of the nanomorphic cell is postulated to be of ~10 μm, a typical size of the living cell (see Table 1.1). In fact, the living cell is a marvelous machine, which, in order to achieve the goal of staying alive, not only acquires, processes, and uses information, but also does it at incredibly low rates of energy consumption in the range of femptowatt to nanowatt. Such levels of power would be a dream target for electronic microsystems.

There is a parallel in thinking about micron-scale integrated systems to that which occurred when integrated circuits technology redefined manufacturing of electronic systems. It is hoped that future research will lead to a similar conceptual leap for the fabrication of systems like the nanomorphic cell.

The studies that follow in this book on limits for the required technologies indicate that a functional micron-scaled system might be feasible. This suggests possible new research directions in extremely scaled microsystems and semiconductor bioelectronics, including, for example, integrated micro-scale energy sources, intelligent microsensor arrays, and very low-energy communication and computation.

References

[1] The International Technology Roadmap for Semiconductors, 2007; <http://www.itrs.net/>

[2] R.K. Cavin, V.V. Zhirnov, D.J.C. Herr Djc, A. Avila, J. Hutchby, Research directions and challenges in nanoelectronics, J. Nanoparticle Res. 8 (2006) 841–858.

[3] R.K. Cavin, V.V. Zhirnov, Morphic architectures: Atomic-level limits, Mater. Res. Symp. Proc. 1067E (2008) B01–02.

[4] K.D. Wise, Integrated sensors, MEMS, and microsystems: Reflections on a fantastic voyage, Sensors and Actuators A 136 (2007) 39–50.

[5] J. Rabaey, J. Ammer, B. Otis, E. Burghardt, Y.H. Chee, N. Pletcher, et al., Ultra-low-power design – The roadmap to disappearing electronics and ambient intelligence, IEEE Circ. Dev. 22 (2006) 23–29.

[6] E.A. Johannessen, L. Wang, S.W.J. Reid, D.R.S. Cumming, J.M. Cooper, Implementation of radiotelemetry in a lab-on-a-pill format, Lab on a Chip 6 (2006) 39–45.

[7] B.W. Cook, S. Lanzisera, K.S.J. Pister, SoC issues for RF smart dust, Proc. IEEE 94 (2006) 1177–1196.

[8] A. Kinkhabwala, P. IH Bastiaens, Spatial aspects of intracellular information processing, Current Opinion in Genetics & Development 20 (2010) 31–40.

[9] A. Wagner, From bit to it: How a complex metabolic network transforms information into living matter, BMC Systems Biology 1 (2007) 33.

[10] A. Danchin, Bacteria as computer making computers, FEMS Microbiol. Rev. 33 (2009) 3–26.

[11] A.L. Koch, What size should a bacterium be? A question of scale, Ann. Rev. Microbiol. 50 (1996) 317–348.

[12] A.M. Makarieva, V.G. Gorshkov, B.-L. Li, S.L. Chown, P.B. Reich, V.M. Gavrilov, Mean mass-specific metabolic rates are strikingly similar across life's major domains: Evidence for life's metabolic optimum, Proc. Natl. Acad. Sci. 105 (2008) 16994–16999.

[13] D. Bray, Protein molecules as computational elements in living cells, Nature 376 (1995) 307–312.

[14] N. Ramakrishnan, U.S. Bhalla, J.J. Tyson, Computing with proteins, Computer 42 (2009) 47–56.

[15] N. Lane, Mitochondria: Key to complexity, in: W. Martin, M. Müller (Eds.), Origing of Mitochondria and Hydrogenosomes, Springer-Verlag, Berlin Heidelberg, 2007.

[16] W.B. Whitman, D.C. Coleman, W.J. Wiebe, Prokaryotes; The unseen majority, Proc. Nat. Acad. Sci. 95 (1998) 6578–6583.

[17] A.M. Makarieva, V.G. Gorshkov, B.-L. Li, S.L. Chown, Size- and temperature-independence of minimum life-supporting metabolic rates, Functional Ecology 20 (2006) 83–96.

[18] A.E. Black, Critical evaluation of energy intake using the Goldberg cut-off for energy intake: basal metabolic rate. A practical guide to its calculation, use and limitation, Intern. J. Obesity 24 (2000) 1119–1130.

[19] H.C. Mond, D. Hunt, J. Vohra, J.G. Sloman, Cardiac Pacing: Memories of a Bygone Era, PACE 31 (2008) 1192–1201.

[20] R.S. Sanders, M.T. Lee, Implantable Pacemakers, Proc. IEEE 84 (1996) 480–486.

[21] R. Fernandes, D.H. Gracias, Toward a miniaturized mechanical surgeon, Materials Today 12 (2009) 14–20.

[22] A. Karargyris, N. Bourbakis, Wireless capsule endoscopy and endoscopic imaging, IEEE Eng. Med. and Biol. 29 (2010) 72–83.

[23] D.C. O'Brien, J.J. Liu, G.E. Faulkner, S. Sivathasan, W.W. Yuan, S. Collins, et al., Design and implementation of optical wireless communications with optically powered smart dust motes, IEEE J. Select. Areas Commun. 27 (2009) 1646–1653.

[24] A.M. Sodagar, K.D. Wise, K. Najafi, a wireless implantable microsystem for multichannel neural recording, IEEE Trans. Microwave Theory and Techn. 57 (2009) 2565–2573.

[25] A. Dufresne, M. Salanoubat, F. Partensky, F. Artiguenave, I.M. Axmann, V. Barbe, et al., Genome sequence of the cyanobacterium Prochlorococcus marinus SS 120, a nearly minimal oxyphototrophic genome, PNAS 100 (2003) 10020–10025.

[26] B.A. Warneke, M.D. Scott, B.S. Leibowitz, L. Zhou, C.L. Bellew, J.A. Chediak, et al., An autonomous 16 mm3 solar-powered node for distributed wireless sensor networks, 1st IEEE Intern. Conf. on Sensors (June 12–14, 2002, Orlando, FL) 2 (2002) 1510–1515.

CHAPTER

Energy in the small: Integrated micro-scale energy sources

2

CHAPTER OUTLINE

2.1 INTRODUCTION

Embedded energy sources are a key enabler for applications with limited or no physical access to external energy supplies. For nanomorphic micron-scale systems, such as integrated analytical microsystems, implantable diagnostics, drug delivery devices, etc., the available volume for on-board energy supplies is very limited. Thus, the capacity of an energy supply, both in terms of energy stored and the rate at which it can deliver energy, can place severe constraints on system design.

In this chapter, fundamental scaling limits for a variety of micron-sized energy sources are investigated, including the galvanic and fuel cells, the supercapacitor, and radioisotope sources. Further, it is sometimes possible to directly harvest energy from the surrounding environment when the delivery of the energy can be either unintentional or directed. In the unintentional category, energy sources might include ambient electromagnetic, solar, vibration, fluid flows, thermal gradients, etc. An overview of the energy available for harvesting from a variety of sources is also provided.

BOX 2.1 MAIN CONCEPT: 'ATOMIC FUEL'

All known sources of energy use the released energy from breaking or forming inter-atomic or intra-atomic bonds. It is important to remember, therefore, that the minimum energy-storing element is the atom or molecule and that energy release is always accompanied by '*burning atoms.*' The total energy available is proportional to the number of energy-releasing atoms, N_{at}:

$$E = \varepsilon N_{at}$$

Where ε is the energy released per atom.

BOX 2.2 AVOGADRO'S NUMBER

According to Avogadro's Law, the number of molecules in one mole of *any substance* is constant. It is called Avogadro's Number N_A:

$$N_A = 6.022\text{x}10^{23}\ mol^{-1}$$

The number of atoms, n_m, in a unit mass of a substance can be found from Avogadro's Number, N_A, and the molar mass, M, which corresponds to the relative atomic mass from the Periodic Table expressed, e.g., in grams.

$$n_m = \frac{N_A}{M}$$

For example, for metallic Zn, M = 65 g/mole and $n_m = 9.26 \times 10^{21}$ at/g

The atomic density, which is the number of atoms in a unit volume of a substance can be found as:

$$n_{at} = \rho \cdot n_m = \frac{\rho N_A}{M}$$

where ρ is the density of the substance (often shown in the Periodic Table).

For metallic Zn, $\rho = 7.14$ g/cm^3 and $n_{at} = 6.61 \times 10^{22}$ at/cm^3.

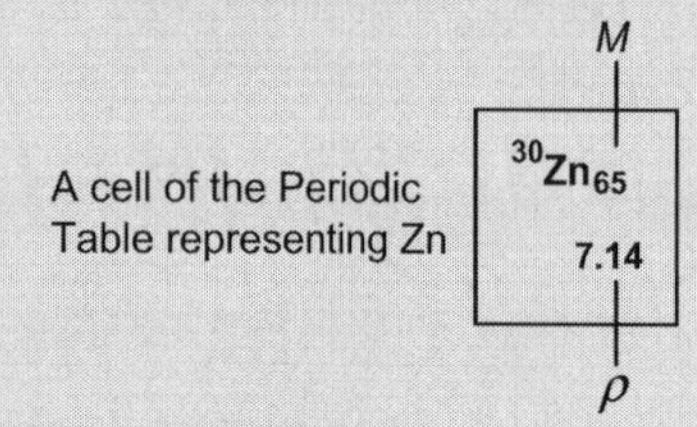

BOX 2.3 GEOMETRIC RELATIONS IN SOLIDS

Atomic density – Number of atoms per unit of volume (e.g. per cm^3):

$$n_{at} = n_{3D} = n$$

Number of atoms per unit of length in an arbitrary direction (e.g. per cm):

$$n_{1D} \sim \sqrt[3]{n_{3D}}$$

Nearest-neighbor distance:

$$l_{a-a} \sim n^{-\frac{1}{3}}$$

Surface concentration – Number of atoms per unit of area in an arbitrary cross-section (e.g. per cm^2):

$$n_s = n_{2D} \sim n^{\frac{2}{3}}$$

2.2 ELECTROCHEMICAL ENERGY: FUNDAMENTALS OF GALVANIC CELLS AND SUPERCAPACITORS

A galvanic cell, in its simplest form, consists of two metal electrodes separated by a layer of electrolyte, which allows for ion transport between the electrodes (Fig. 2.2.1). The right (negative) metal electrode in Figure 2.2.1 loses its atoms, e.g. Li or Zn, which are converted into ions (through an electrochemical reaction), e.g. Li^+, Zn^{2+}, Al^{3+}, as they go into solution. The electrode becomes negatively charged, due to excess electrons that flow through the external connection. Thus, in the galvanic cell, '*atomic fuel*' is consumed at the negative electrode to produce electricity: For every 1–2 electrons that flow through the external circuit, a metal atom must go into the electrolyte solution as

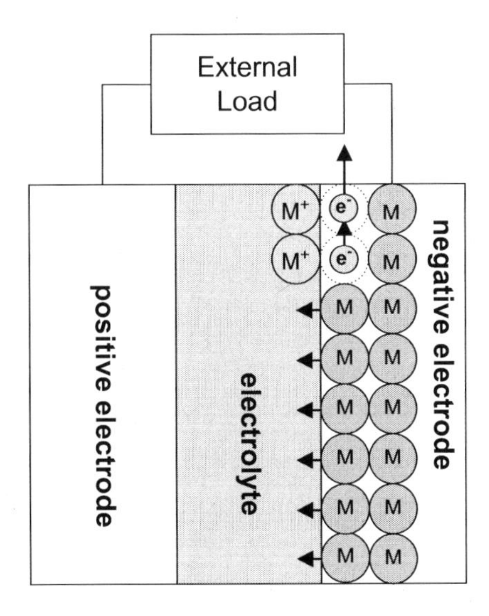

FIGURE 2.2.1

Schematic diagram of a galvanic cell

Table 2.2.1 Electrochemical energy density metrics for several metals

Negative electrode[1] reaction	Electrode potential (V)*	N_{el}	Characteristic energy density metrics**		Reference
			J/cm³	J/g	
Al^{3+}/Al	−1.68	3	4.86E + 04	1.80E + 04	[1-3]
Mg^{2+}/Mg	−2.38	2	3.33E + 04	1.91E + 04	[3]
Mn^{2+}/Mn	−1.18	2	3.07E + 04	4.13E + 03	[1]
Ti^{2+}/Ti	−1.63	2	2.94E + 04	6.54E + 03	[4]
Li^{+}/Li	−3.05	1	2.23E + 04	4.20E + 04	Commercial lithium batteries
Zn^{2+}/Zn	−0.76	2	1.61E + 04	2.25E + 03	Commercial zinc-carbon batteries
Fe^{2+}/Fe	−0.44	2	1.19E + 04	1.51E + 03	Ni-Fe
Cd^{2+}/Cd	−0.40	2	5.95E + 03	6.88E + 02	Commercial nickel-cadmium batteries

*Electrode potential vs. the standard hydrogen electrode.
**Calculated using formulas in Box 2.2.

a positively charged ion M^+. When the supply of the metal fuel atoms is exhausted, the galvanic cell can no longer provide energy.

2.2.1 Energy stored in the galvanic cell

Because the typical chemical bonding energy per electron is on the order of a few eV, the typical potential difference V produced by such a system is ~ 1 Volt. Thus, one atom of the atomic fuel produces energy $\varepsilon \sim eV$ and the total stored energy E_{stored} can be estimated as

$$E_{stored} = \varepsilon N_{at} = eN_{el} \cdot N_{at} \cdot V \sim eN_{at} \cdot V \tag{2.2.1}$$

where e is the charge on an electron, N_{el} is the number of electrons released per atom, and N_{at} is the number of atoms comprising the metal electrode.

The upper bound for the energy in an electrochemical source can be estimated by using the fact that the number of molecules (or atoms) in one mole of matter is given by Avogadro's Number, $N_A = 6.02 \times 10^{23}$ at/mol, and that the atomic density in all solids, n_{at}, varies from 10^{22} to 10^{23} at/cm³:

$$E_{\max} \sim e \cdot N_A \cdot (1V) = 1.6 \times 10^{-19} \cdot 6.02 \times 10^{23} \sim 10^5 \frac{J}{mole} \tag{2.2.2a}$$

or

$$E_{\max} \sim e \cdot n_{at} \cdot (1V) = 1.6 \times 10^{-19} \cdot 10^{23} \sim 10^4 \frac{J}{cm^3} \tag{2.2.2b}$$

In Table 2.2.1, the gravimetric (J/g) and volumetric (J/cm³) electrochemical energy densities for several metal electrodes are characterized.

[1]The electrochemical community refers to the 'negative electrode' as the 'anode' and the positive electrode as the 'cathode.' However, physicists and electrical engineers use an inverted notation. Because of conflicting conventions for anode and cathode usage and because this paper primarily targets an audience beyond the professional electrochemical community, the terms 'positive electrode' and 'negative electrode' are used.

Finally, from (2.2.2b), if the entire volume of a 10 μm-sized system is filled with the atomic fuel, a maximum of about 10^{-5} Joule would be available.

The energy output is limited by the *number of atoms* available for conversion. When all atoms from the negative electrode are converted into ions and the source is depleted, then continued use requires one of the following actions:

1. Replace the cell.
2. Recharge the cell by applying external electric energy and reversing the electrochemical reactions (converting ions back to atoms).
3. 'Refill' the cell by replacing the electrode material. Conceptually, this is how fuel cells operate. Some examples will be discussed in Section 2.2.4.

Of course, in reality the 10 μm^3 volume must be shared with other components, and indeed a more accurate account for use of the volume would also include all essential components of the galvanic cell, i.e. the anode, cathode, electrolyte, and encapsulation.

2.2.2 Power delivery by a galvanic cell

The power delivery by a galvanic cell is

$$P = \frac{E_s}{t_s} \tag{2.2.4}$$

where t_s is the energy release time and E_s is the energy released by the surface atoms, which are in contact with the electrolyte and available for electrochemical reaction. Using an approximate relationship between the number of atoms in the volume n_{at} and atoms on the surface n_s (see Box 2.3), obtain:

$$E_s = \varepsilon \cdot n_s \approx \varepsilon \cdot n_{at}^{2/3} \tag{2.2.5a}$$

The energy release time t_s in the limit (e.g. very small resistance of the electrolyte) depends on the time constants of the electrode redox reaction (i.e. ion formation), which are in the range of 10^{-2}–10^{-4} s [23].

The upper bound for the power of an electrochemical source can now be estimated as

$$P_{\max} \sim \frac{e}{t_s} \cdot E_{at} \cdot n_{at}^{2/3} \sim \frac{1.6 \cdot 10^{-19}}{10^{-4}} \cdot 1 \cdot (10^{23})^{\frac{2}{3}} \sim 1 \frac{W}{cm^2} \tag{2.2.5b}$$

assuming $\varepsilon \sim 1$ eV, $n_{at} \sim 10^{23}$ at/cm^3 and $t_s \sim 10^{-4}$ s.

Finally, from (2.2.5b) for a 10 μm-sized galvanic cell, the upper bound of attainable power is ~ 10^{-6} Watt.

2.2.3 Current status of miniature galvanic cells

Remarkable progress on miniature battery technology has been reported. For example, Cymbet Corp. has a rechargeable thin-film Li battery with package dimensions of 5.00 × 5.00 × 1.05 mm [5]. The output voltage is 3.8 V and the battery capacity is 12 μAh. This battery targets such applications as non-volatile SRAM, localized power sources for microcontrollers, etc. It should be noted, that while Li-based batteries are the primary choice for powering portable devices (because of their energy density), other kinds of batteries may better fit the constraints for extremely scaled bioelectronic systems:

1. Li appears to be one of the best materials from the point of view of energy storage; it has the largest standard electrode potential (–3.05 V) and it is the lightest solid material ($\rho = 0.53$ g/cm^3). Therefore, Li-based batteries provide the highest gravimetric energy density (J/g), as indicated in Table 2.2.1, which is an important consideration for lighter-weight portable applications and for transportation. However, for applications where space is the primary constraint, other metals such as Al, Mg, Mn, and Ti outperform Li in volumetric energy density. In terms of volumetric proficiency, e.g., Zn (the most popularly used primary battery at present) is comparable to Li.
2. The operating voltage, V, is above 3 Volts in Li batteries; this is attractive from the point of view of maximizing the total stored energy E_{stored}, which, according to (2.2.1), is directly proportional to V. On the other hand, the energy dissipation in the load for electronic circuits is proportional to V^2. For example, as is discussed in Chapter 3, Box 3.6, operation of electronic devices always involves charging and discharging of an equivalent capacitor C; the corresponding energy dissipated in an elementary switching event is

$$E_{sw} = CV^2 \tag{2.2.6}$$

The total number of the operations the system can perform with a given energy supply (e.g. elementary binary switching events) is according to (2.2.1) and (2.2.6):

$$N_{sw} = \frac{E_{stored}}{E_{sw}} \sim \frac{1}{V} \tag{2.2.7}$$

Thus, the total number of switching events decreases as voltage increases (Fig. 2.2.2). For these reasons the operating voltage should be below 1 V (for more on this see the following sections). If Li

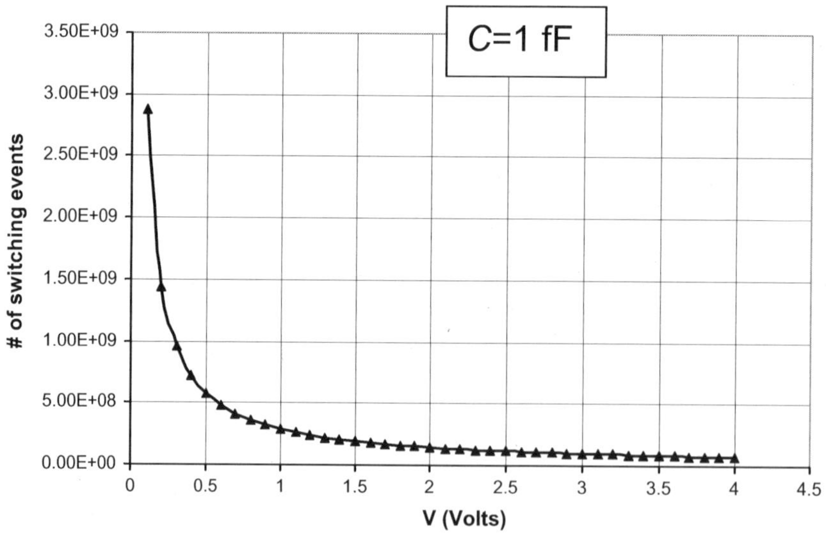

FIGURE 2.2.2

The total number of elementary switching events for fixed total energy as a function of operational voltage

sources were used in the nanomorphic system, they would require voltage conversion, consume additional volume, and dissipate additional energy due to energy conditioning.

3. Encapsulation may be the most important issue for very small batteries, especially when electrode materials that are not compatible with water are used. Lithium violently reacts with water and the need for a package (case) may impose a limit to practical miniaturization [6–8]. To address this problem, caseless microbatteries have been proposed for bio-implantable applications [6]. In fact, such caseless microbatteries consist only of two electrodes immersed in physiological fluids such as the subcutaneous interstitial fluid, blood, serum, etc. Polymer-coated Zn and Ag/AgCl could be used as negative and positive electrodes respectively, since these materials are generally regarded as potentially harmless and could therefore be used for in vivo galvanic microcells [6]. An experimental cell Zn|serum|Ag/AgCl made with a fine Zn fiber anode was shown to operate at 1V and 13 μA/cm^2 for about 2 weeks [7]. Extreme cell miniaturization is currently in progress and preliminary results on experiments with galvanic cells having a 10 μm-disk Zn anode have been reported [8]. As another example, a miniature Zn-Pt galvanic cell utilizing gastric fluid in the stomach as the electrolyte was described in [31]. The size of the fabricated battery was 10 mm × 8 mm × 4 mm. In experiments using simulated gastric fluid (80 mM HCl, 34 mM NaCl, and 10 μM digestive enzyme pepsin) as the electrolyte, this battery generated 1 mW of power.

2.2.4 Miniature biofuel cells

The primary attraction of fuel cells is that their energy capacity is not limited as long as energy (in chemical form) is supplied externally. For bioelectronic applications, e.g., it is attractive to use a small amount of energy stored in a biological organism to power, for instance, the nanomorphic microsystem [32]. One important 'biofuel' is glucose – $C_6H_{12}O_6$. The glucose–O_2 biofuel cell is a promising candidate for micropower source [8–11,33,34]. The two reactants (glucose and O_2) are blood-supplied and are present in most tissues.

Another important 'molecular fuel' in biosystems is adenosine 5'-triphosphate (ATP) – $C_{10}H_{16}N_5O_{13}P_3$. The ATP energy storage density is ~6.6 × 10^4 J/kg or ~3 × 10^4 J/mole [12,13]. This corresponds to ~0.3 eV per ATP molecule, which is comparable to ~ 1 eV in the galvanic cell. However, in biosystems, the ATP is dissolved to a typical concentration of 1–10 mM (10^{-6}–10^{-5} mole/cm^3) [12,13], and thus the maximum stored energy in ATP is reduced to ~0.03–0.3 J/cm^3. One possible approach to boosting the energy density of micro-biofuel cells is to use a microfluidic 'concentrator' [13].

The upper bound for the power of a fuel cell harvesting power from a human body can now be estimated in the same way as (2.2.5) using parameters in Table 2.2.2:

$$P_{max} \sim \frac{e}{t_s} \cdot \varepsilon \cdot n^{2/3} \sim \frac{1.6 \cdot 10^{-19}}{10^{-4}} \cdot 30 \cdot \left(3 \cdot 10^{18}\right)^{\frac{2}{3}} \sim 0.1 \frac{W}{cm^2} \tag{2.2.8}$$

(assuming glucose fuel with ε ~ 30 eV, n ~ 3 × 10^{18} molecule/cm^3, and t_s ~ 10^{-4} s).

Similar calculations for ATP result in ~0.002 W/cm^2. Finally, for a 10 μm-sized biofuel cell, the upper bound of attainable power is ~ 10^{-7} Watt for glucose and ~ 10^{-9} Watt for ATP.

Biofuel cells are currently an active area of research and recent reviews for the field are contained in [33] and [34]. An example of a very small caseless cell, consisting of two bio-electro-catalyst-coated

Table 2.2.2 Characteristics of biofuels available in living systems

	Glucose	ATP
ε, eV/molecule	~30	~0.3
Concentration in human blood, mM	~5	10
n, molecule/cm^3	~3 × 10^{18}	~6 × 10^{18}
Max. power density, W/cm^2	0.1	0.002
Max. power in nanomorphic cell, W	10^{-7}	10^{-9}

carbon fibers 7 μm in diameter and 2 cm in length, gave a 0.26 mm^2 footprint and a 0.0026 mm^3 volume [10]. This cell demonstrated continuous generation of 4.4 μW in a physiological, glucose-enriched buffer solution (30 mM glucose). Operation of this miniature biofuel cell implanted in a living plant (grape) has also been demonstrated, generating 2.4 μW [11].

2.2.5 Remarks on biocompatibility

With caseless microbatteries, consisting of two electrodes immersed in physiological fluids, dilute quantities of metal are injected into the body – a potentially undesirable event. However, the total mass of metal in the 10-μm cube does not exceed 10 ng, which is many orders of magnitude less than typical dietary intakes [14,15] or even regulated drinking water concentrations [16], as shown in Table 2.2.3. The injected nanodoses of elements and compounds would be many orders of magnitude below their 'toxicity' or 'nutrition' levels. Therefore, the contribution of micron-scale galvanic cells should be

Table 2.2.3 Electrochemical energy density metrics along with typical dietary and regulated drinking water concentrations for several metals

Negative electrode reaction	Electrode potential (V)	Characteristic energy density (J/cm^3)	Max. total mass in 10 μm cube	Typical dietary intakes	US EPA drinking water regulations [14]
Al^{3+}/Al	−1.68	4.86E + 04	3 × 10^{-9} g	n/a	5 × 10^{-5} g/L
Mg^{2+}/Mg	−2.38	3.33E + 04	2 × 10^{-9} g	0.3 g/day [14]	~10^{-2} mg/L*
Mn^{2+}/Mn	−1.18	3.07E + 04	7 × 10^{-9} g	2 × 10^{-3} g/day [14]	5 × 10^{-5} g/L
Zn^{2+}/Zn	−0.76	1.61E + 04	7 × 10^{-9} g	10^{-3} g/day [14]	5 × 10^{-3} g/L
Fe^{2+}/Fe	−0.44	1.19E + 04	8 × 10^{-9} g	10^{-3} g/day [14]	3 × 10^{-4} g/L
Cd^{2+}/Cd	−0.40	5.95E + 03	9 × 10^{-9} g	17 × 10^{-6} g/day [15]	5 × 10^{-6} g/L

Not regulated by EPA – the number in the table represents a typical value.

well below limits that would affect human health. (Biocompatibility estimates depend obviously on the number of devices.)

In this section, the limits of energy storage for several different material systems for galvanic and fuel cells have been examined in the context of their applicability as energy sources for the nanomorphic cell. Although the amount of energy available is quite small in either case, this study suggests that the idea of using galvanic cells as an energy source for micron-scale systems is not out of question. The ultimate usefulness of galvanic cells in this context depends strongly on the volume and energy requirements and trade-offs for sensing, computation, and communication that are discussed in subsequent chapters.

2.2.6 Miniature supercapacitors

The micro-scale electrochemical sources described in the previous sections have power output in the μW range, which could be sufficient for some applications (e.g. sensing, logic, control), but too small for others (e.g. signal transmission). Integration of a micro-scale battery with a micro-scale supercapacitor may help boost the power output of the system [10,11].

A schematic diagram of a supercapacitor is shown in Figure 2.2.3. One plate of the capacitor is formed by the metal electrode, while the electrolyte forms the second plate. In order to make external contact, a second electrode is needed. The two-electrode structure shown in Figure 2.2.3 is in effect two capacitors connected in series (a capacitance exists at each of the electrode/electrolyte interfaces).

If the negative electrode/electrolyte capacitance is C_- and the positive electrode forms a capacitance C_+, the resulting capacitance of the two serial capacitances is

$$C = \frac{C_- C_+}{C_- + C_+} = \frac{C_-}{\frac{C_-}{C_+} + 1} \qquad (2.2.9)$$

If $C_- = C_+$ (symmetric capacitor), the resulting capacitance $C = C_-/2$. If $C_- << C_+$ (asymmetric capacitor), $C \approx C_-$.

There are two energy storage mechanisms in supercapacitors. The first mechanism results from the formation of an electrical double-layer at the electrode/electrolyte interface. The second energy storage mechanism is due to the voltage-dependent fast electrochemical (faradaic) reactions occurring at the electrode surface between electrode atoms and ions of the electrolyte. The double-layer capacitance always exists at the metal/electrolyte interface. In addition, the faradaic capacitance effect can occur with certain electrode materials. The two mechanisms each have their characteristic capacitances.

The capacitance of a double-layer supercapacitor can be estimated from a standard equation [17]:

$$C = \frac{\varepsilon_0 K A}{d} \qquad (2.2.10)$$

where $\varepsilon_0 = 8.85 \times 10^{-14}$ F/cm is the permittivity of free space, K is the relative dielectric constant of the interpolate layer, d is the thickness of the interpolate layer (Fig. 2.2.3), and A is the surface area of the capacitor plates.

The basic idea in supercapacitors is to maximize C by creating the minimum possible d in (2.2.10) due to the use of the electrical double-layer at the metal–electrolyte interface (Fig. 2.2.3). For concentrated aqueous electrolytes, $d = 0.5$–1 nm [17]. By using this value and the dielectric constant

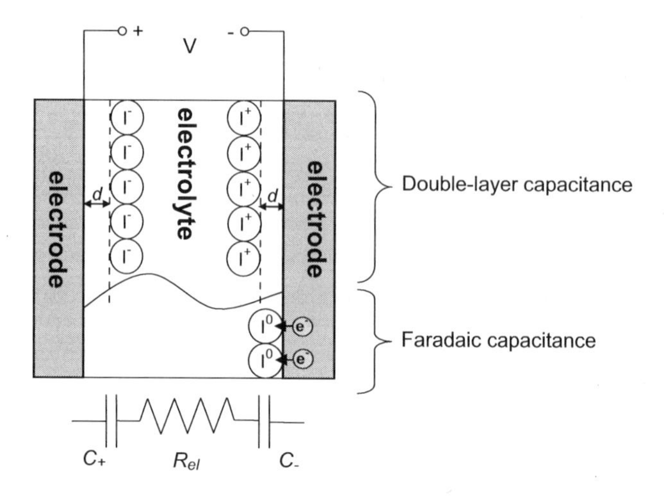

FIGURE 2.2.3

A schematic diagram of a supercapacitor. The two parallel energy storage mechanisms in supercapacitors are shown: double-layer capacitance and faradaic capacitance. In general, the electrode capacitances C_+ and C_- are the sum of both. The double-layer capacitance always exists at the metal/electrolyte interface. In addition, the faradaic capacitance effects can occur with certain electrode materials

inside the double-layer, $K \sim 10$ for aqueous electrolytes [17], (2.2.9) gives the specific capacitance (the capacitance per unit area, $c_s = C/A$) as $C_s \sim 10$–$20\ \mu F/cm^2$.

The faradaic capacitance (pseudocapacitance) is defined as $C_f = \frac{q}{V}$, where q is the charge transferred during the electrochemical reaction. The maximum value of q is limited by the surface density of atoms n_s (which in all solids varies from ~10^{14} at/cm^2 to ~10^{15} at/cm^2):

$$q_{\max} \sim e \cdot n_s \sim 1.6 \times 10^{-19} \cdot 10^{15} \sim 10^{-4} \frac{C}{cm^2} \tag{2.2.11}$$

For $V \sim 1$ V, the faradaic capacitance per unit area is $C_f \sim 100\ \mu F/cm^2$ or about 10x larger than the double-layer capacitance.

The total capacitance can be further increased by increasing surface area using porous electrode materials with high specific surface a_s (total area per unit mass). Different forms of activated carbon offer the highest specific surface – on the order of 1000 m^2/g [17–21]. The upper bound for the specific surface of carbon was estimated in [18] based on atomic packing considerations of graphene sheets in graphite (Fig. 2.2.4). Since the size of the pores cannot be less than the distance between the neighbor sheets, $\delta = 0.335$ nm, this distance gives us a lower bound for the pore size (and therefore the upper bound of porosity). Let the total surface of one graphene sheet be $A = 2lw$ (both sides result in the factor of two), the volume $v = lw\delta$, and the mass v ($\rho = 2.267$ g/cm^3 for graphite). The specific area of such 'ultimately porous' carbon is

$$a_s = \frac{A}{m} = \frac{2lw}{\rho v} = \frac{2lw}{\rho lw\delta} = \frac{2}{\rho\delta} \tag{2.2.12a}$$

or numerically:

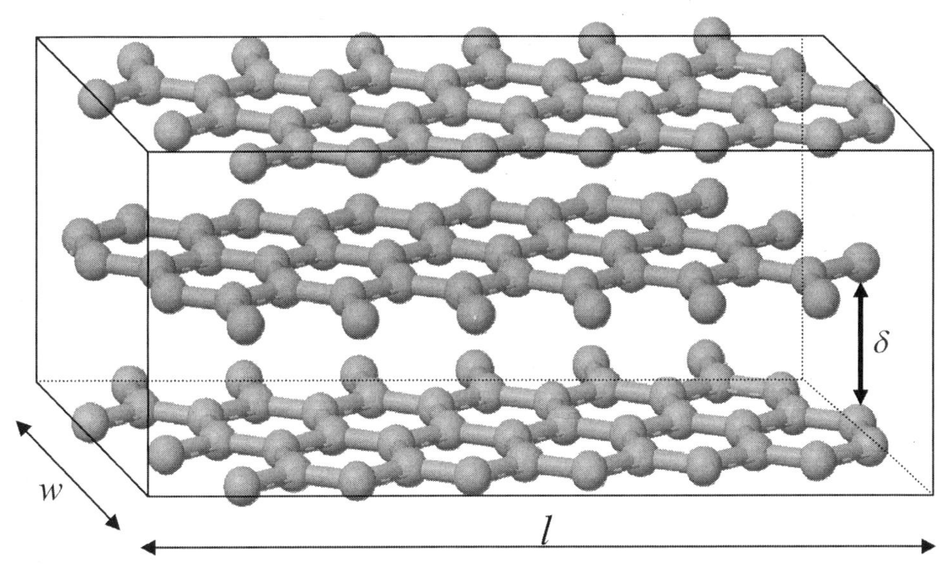

FIGURE 2.2.4

Atomic structure of graphite used to estimate the upper bound for porosity in carbon materials. The size of an 'ultimate' pore cannot be less than the distance between neighbor graphene sheets, δ

$$a_s = \frac{2}{2.267 \cdot 3.35 \times 10^{-8}} = 2.6 \times 10^7 \frac{cm^2}{g} = 2600 \frac{m^2}{g} \tag{2.2.12b}$$

This simple estimate is consistent with rigorous calculations of, e.g., the specific surface area of the 'ropes' of parallel single-wall nanotubes in the limit of ultimately small nanotube diameter [21].

A widely used characteristic number is the gravimetric specific capacitance (per unit mass):

$$c_m = a_s \cdot c_s \tag{2.2.13}$$

Assuming that the value of a_s in (2.2.12b) is an approximate upper bound for all material systems and using the above estimates $c_s \sim 10\ \mu F/cm^2$ (double-layer) and $c_s \sim 100\ \mu F/cm^2$ (faradaic) for specific capacitances, the approximate upper bound for the gravimetric specific capacitances: $c_m \sim 300$ F/g (double-layer) and $c_m \sim 3000$ F/g (faradaic) is obtained. Table 2.2.4 contains specific capacitances of several materials systems and compares these with the upper bound numbers.

For a cubic electrode of size l, the total surface area is

$$A = m \cdot a_s = \frac{2\rho l^3}{\rho \delta} = \frac{2l^3}{\delta} \tag{2.2.14}$$

If $l = 10\ \mu m$, then $A \sim 0.06\ cm^2$ and the total capacitance of a double-layer supercapacitor is $\sim (5 - 10)\frac{\mu F}{cm^2} \times 0.06\ cm^2 = 0.3 - 0.6\ \mu F$(upper bound). The specific capacitance of the faradaic capacitors is roughly 10–20 times larger, and therefore ~10 μF is the upper bound for the 10-μm size.

The energy stored in the capacitor is

$$E = \frac{CV^2}{2} \tag{2.2.15}$$

Table 2.2.4 Specific capacitances of several materials systems

Double-layer capacitors	c_m, F/g	Ref.
Upper bound	**~300**	**[18] and this work**
Activated carbons/aq-H_2SO_2	200	[19]
Activated carbon/NaCl-gelatine	80	[26]
Faradaic capacitors		
Upper bound	**~3000**	**this work**
RuO_2/aq-H_2SO_4	900	[30]
$MnO_2 \cdot nH_2O$/aq-NaCl	200	[24]
$MnFe_2O_4$/aq-NaCl	100	[25]

While it might be desirable to increase the voltage V for larger stored energy, the maximum voltage V_{max} of the electrochemical double-layer capacitors is fundamentally limited by the electrolysis threshold voltage. For aqueous electrolytes, $V_{max} = 1.2$ Volts, and thus the maximum energy stored in the capacitor $E_{\max} \sim 0.5 \cdot 0.6\ \mu F \cdot (1.2V)^2 = 4 \times 10^{-7} J$(double-layer) and $\sim 10^{-5} J$ (faradaic).

The capacitor voltage can be increased to some extent by using organic electrolytes ($V_{max} \sim 2.5$ V) or ionic liquids – organic molten salts ($V_{max} \sim 4$ V). In the latter case, $E_{\max} \sim 5 \cdot 10^{-6} J$(double-layer) and $\sim 10^{-4} J$ (faradaic).

The power delivery by discharge of a capacitor is

$$P = \frac{E}{t} \tag{2.2.16}$$

where t is the energy release time, which depends on the time constants of the electrode reactions (e.g. the formation/relaxation of double-barrier or redox reactions) t_{el}, and the capacitor discharge time t_C:

$$t = t_{el} + t_C \tag{2.2.17}$$

The electrode reaction time is $t_{el} \sim 10^{-8}$ s for the double-barrier formation process and $t_{el} \sim 10^{-4}$s for faster redox reactions [23]. The capacitor discharge time $t_C \sim 2RC$ (double-layer capacitor, roughly 90% V-discharge, 99% E-discharge), where R is the resistance of the electrolyte, which can be calculated using the conventional formula:

$$R = \frac{d}{\sigma A} \tag{2.2.18}$$

In (2.2.18), σ is the conductivity of the electrolyte, $A = l^2$ is the planar electrode area, and d is the electrolyte layer thickness. The electrolyte layer thickness must be at least larger than 2 × the double-layer thickness; therefore, following [18], assume $d = 2$ nm for the hypothetical best case scenario. For $d = 10$ μm and $\sigma \sim 0.7$ S/cm (concentrated aqueous electrolyte), the resulting resistance $R \sim 0.3\ \Omega$. For the ionic liquids $\sigma \sim 0.01$ S/cm, which results in $R \sim 20\ \Omega$.

Characteristic parameters driving the performance of supercapacitors are given in Table 2.2.5, along with estimates using (2.2.15)–(2.2.18) for the energy and power delivery by a 10 μm-sized supercapacitor. Note that the upper bound for power is ~ 1 W.

Table 2.2.5 Estimated parameters of near-ideal 10 μm-sized double-layer and faradaic supercapacitors

Capacitor	Double-layer		Faradaic	
Electrolyte	**Aqueous**	**Ionic**	**Aqueous**	**Ionic**
V_{max} (Volt)	~1.2	~4	~1.2	~4
C_{max} (F)	~6 x 10^{-7}	~6 x 10^{-7}	~10^{-5}	~10^{-5}
R_{min} (Ω)	~0.3	~20	~0.3	~20
t_{el} (s)	~10^{-8}	~10^{-8}	~10^{-4}	~10^{-4}
t_C (s)	~4 x 10^{-7}	~2 x 10^{-5}	~6 x 10^{-6}	~4 x 10^{-4}
t (s)	~4 x 10^{-7}	~2 x 10^{-5}	~10^{-4}	~5 x 10^{-4}
E_{max} (J)	~4 x 10^{-7}	~5 x 10^{-6}	~7 x 10^{-6}	~8 x 10^{-5}
P_{max} (W)	~1	~0.2	~0.07	~0.16

Miniature supercapacitors: Discussion on the status and research needs

As seen from Table 2.2.5, the double-layer supercapacitors have higher potential for power delivery than the faradaic capacitors. Studies of new mechanisms of electrode–electrolyte interactions could help to improve the achievable capacitance. For example, it was recently shown, using porous carbon materials with a very small pore size (< 1 nm), that a 50% increase in the double-layer capacitance could result [21]. The authors [21] suggest that this effect is due to the decrease of the effective double-layer thickness d in (2.2.10), which is determined by the size of the pore.

It should be noted, however, that the double-layer capacitors achieve their highest power performance in strong electrolytes, such as sulfuric acid. This raises encapsulation issues and thus makes scaling to small dimensions difficult. For maximum voltage, organic molten salts are needed and must be encapsulated. The need for encapsulation may impose a practical scaling limit on double-layer capacitors.

For autonomous bioelectronic microsystems, it is desirable to minimize or eliminate the encapsulation issue (see the related discussion in the microbatteries section above). In principle, it might be possible to make a caseless supercapacitor which uses the physiological fluids (e.g. NaCl-based) as the electrolyte [20]. Research to explore new electrode materials with supercapacitance behavior in NaCl electrolyte is therefore needed. It appears that for caseless applications, the faradaic supercapacitors are more suitable. Mn-based materials are promising for these applications. For example, hydrated amorphous or nanocrystalline manganese oxide, $MnO_2 \cdot H_2O$, exhibits capacitances ~200 F/g in NaCl solutions [24]. Also, $MnFe_2O_4$ was found to exhibit a large faradaic capacitance of $>$ 100 F/g in NaCl electrolyte and high-power delivering capabilities of $>$10 kW/kg [25].

NaCl-contained biopolymer solid electrolytes could have a potential for use in biocompatible double-layer supercapacitors [26]. Such electrolytes have an ionic conductivity of up to 0.1 S/cm [26]. Supercapacitors made of Black Pearl Carbon electrodes and NaCl-doped electrolytes have demonstrated a specific capacitance ~80 F/g [26].

Conductive polymers are also attractive materials for faradaic micro-supercapacitors. They have specific capacitance of 100–300 F/g and can in principle be synthesized accurately to the sub-micron scale by electrochemical methods [27]. Both electrodes and electrolytes can be made from the conductive polymers. For example, in [28], all-solid polymer flexible miniature supercapacitors were fabricated using polypyrrole (PPy) as electrodes and solidified polyvinyl alcohol (PVA) as electrolyte. The lateral dimensions of the capacitor were 2×2 mm and the thickness was much less than 1 mm. The capacitor structure could be deformed, rolled up, etc., without changes in performance. The structure had a capacitance of several mF and was tested at a maximum voltage of 0.6 V. In principle, such an all-solid polymer capacitor could have better scalability due to relaxed encapsulation constraints.

It should also be noted that embedded miniature supercapacitors would find many applications in semiconductor microelectronics. For example, in order to respond to the power integrity problem in microprocessors, integrated thin-film high capacitances are needed [29]. Currently existing supercapacitors are not available in a planar format suitable for on-chip integration. The potential for integrated miniature supercapacitors for microprocessors was recently discussed in [29].

In conclusion, embedded miniature supercapacitors appear to be a potentially feasible power source, whose estimated upper bound for power lies between 0.1 and 1 W.

2.3 ENERGY FROM RADIOISOTOPES

2.3.1 Radioisotope energy sources

Electrochemical sources described in previous sections have energy output ~1 eV/atom, which is related to the energy of inter-atomic bonds. Thus, the maximum energy stored in, e.g., a 10 μm-sized box is $\sim 10^{-5}$ J. In principle, the 'intra-atomic' energy (i.e. of nuclear bonds) is much higher and therefore its utilization seems very attractive for embedded micro-power sources in size-constrained systems [35–37].

The energy of radionuclides is released in energetic particles, typically α- (He ions), β- (electrons), and γ- (electromagnetic radiation) particles. α- and β-emission can in principle be utilized in energy sources. Several examples of α- and β-radioisotopes are given in Table 2.3.1.

The energy release by radionuclides can be calculated using the radioactive decay formula:

$$E(t) = E_{at} \cdot N(t) = E_{at} N_0 \exp\left(-\frac{t}{\tau}\right) \quad (2.3.1)$$

where N_0 is the initial number of the atoms, $N(t)$ is the number of atoms that have not released an energetic particle by the time t, τ is the 'mean life time' of a radioactive atom, and ϵ is the average energy of the particle released by a radioactive atom. For α-emission (discrete energy spectrum, Fig. 2.3.1a), $\varepsilon \approx E_{max}$, while for β-emission (continuous energy spectrum, Fig. 2.3.1b), the average energy of electrons is approximately 1/3 of the maximum energy E_{max}[39].

Another common characteristic time of radioactive reactions is the radionuclide half-life $t_{1/2}$, which is related to the mean life time τ as

$$t_{1/2} = \tau \ln 2 \quad (2.3.2)$$

The total energy released by the radioactive sources is

$$E = \varepsilon N_0 \quad (2.3.3)$$

The average power delivery by a single-atom radioactive reaction is ϵ/t and the power of N radioactive atoms is

Table 2.3.1 Characteristic parameters of several radioisotopes

Radioisotope	(1)E_{max}, eV	(2)τ (s)	(3)L (E_{max}), μm [5,6]	(4)J/cm^3	(5)W/cm^3	Comment
(α) ^{210}Po	5.4×10^6	1.7×10^7	~26	2.3×10^{10}	1.4×10^3	Has been investigated as a light-weight energy source for space applications
(α) ^{238} Pu	5.5×10^6	4.0×10^9	~27	4.4×10^{10}	11	Early use in pacemakers (*Medtronic*, *Numec Corp*. [38])
(β) ^{3}H	1.86×10^4	5.6×10^8	~4	4.9×10^4	8.8×10^{-5}	Used to produce light in self-illuminating watches
(β) ^{63}Ni	6.69×10^4	4.6×10^9	~34	2.3×10^8	5.1×10^{-2}	Demonstrated [35,45–47]
(β) ^{147}Pm	2.24×10^5	3.3×10^5	~300	3.5×10^8	1.1×10^3	Early use in Betacell 400 pacemakers [4] Used in QynCell batteries

(1)*Maximum energy of an energetic particle.*
(2)*Mean life time of a radioactive atom.*
(3)*Stopping range in silicon.*
(4)*Total energy stored in the radioactive source.*
(5)*Total power released by the radioactive source.*

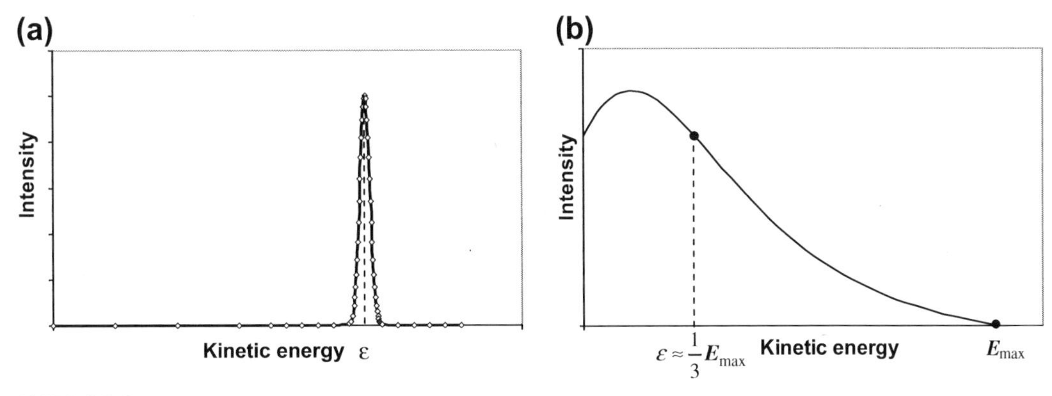

FIGURE 2.3.1

Energy spectra of: (a) α-particles (immediately after decay) and (b) β-decay electrons

$$P = \frac{\varepsilon N_0}{\tau} \tag{2.3.4}$$

The characteristic parameters of several radionuclides, including their energies and power metrics, are given in Table 2.3.1. It would seem that to maximize energy output, the radionuclide with the most energetic particles would need to be used. However, there is a severe constraint on the minimum size of the radioisotope energy sources arising from the ability of matter to absorb radiation. The energy released by radionuclides needs to be captured and converted into a usable form of energy, which can be achieved by the materials adjacent to the radioisotope source. All existing schemes of energy capture/conversion are based on the interaction of the energetic particles with the absorbing matter. Products of this interaction are excess charge, electromagnetic radiation, and heat.

The absorbing matter is characterized by its 'stopping power.' The distance L to the point where the energetic particle has lost all its energy is called the range and it presents a characteristic minimum size for the radioisotope energy source. Stopping ranges in Si for α- and β-particles of different energies are shown in Figure 2.3.2. The stopping range can be approximated by a power function:

$$L(\epsilon) \approx b\varepsilon^k \tag{2.3.5}$$

where b and k are constants, which depend on the type and energy of the energetic particle and on the absorbing material.

The size of a radioisotope energy source (lower bound) is a sum of the 'fuel compartment' size r and the absorber stopping thickness L (given by (2.3.5)), i.e. $S = r + L$ (Fig. 2.3.3a). The size of the 'fuel compartment' determines the maximum energy storage, $E = \varepsilon N_0$, and power delivery of the source. This size depends on the total number of fuel atoms in the compartment's volume ($N_0 = v \cdot n_{at}$):

$$r \sim v^{\frac{1}{3}} = \left(\frac{E}{\varepsilon n_{at}}\right)^{\frac{1}{3}} = \left(\frac{P\tau}{\varepsilon n_{at}}\right)^{\frac{1}{3}} \tag{2.3.6}$$

where E is the total energy stored by the radioisotope source and n_{at} is the atomic density of the radioisotope. Thus, the lower bound on the size of a radioisotope energy source is

$$S_{\min} = b\epsilon^k + \left(\frac{P\tau}{\varepsilon n_{at}}\right)^{\frac{1}{3}} \tag{2.3.7}$$

The maximum power of radioisotope energy sources as a function of total size is shown in Figure 2.3.3b.

2.3.2 Energy conversion

The two large classes of the radioisotopic energy converters are thermal and non-thermal [42].

In principle, conversion of the particle energy into heat allows for ~100% energy collection. However, in conversion of the thermal energy into useful work, the upper bound of the useable output energy is given by the Carnot efficiency limit:

$$E_{out} = \left(1 - \frac{T_c}{T_h}\right) \cdot E_{ab} = \eta E_{ab} \tag{2.3.8}$$

where E_{ab} is the absorbed energy.

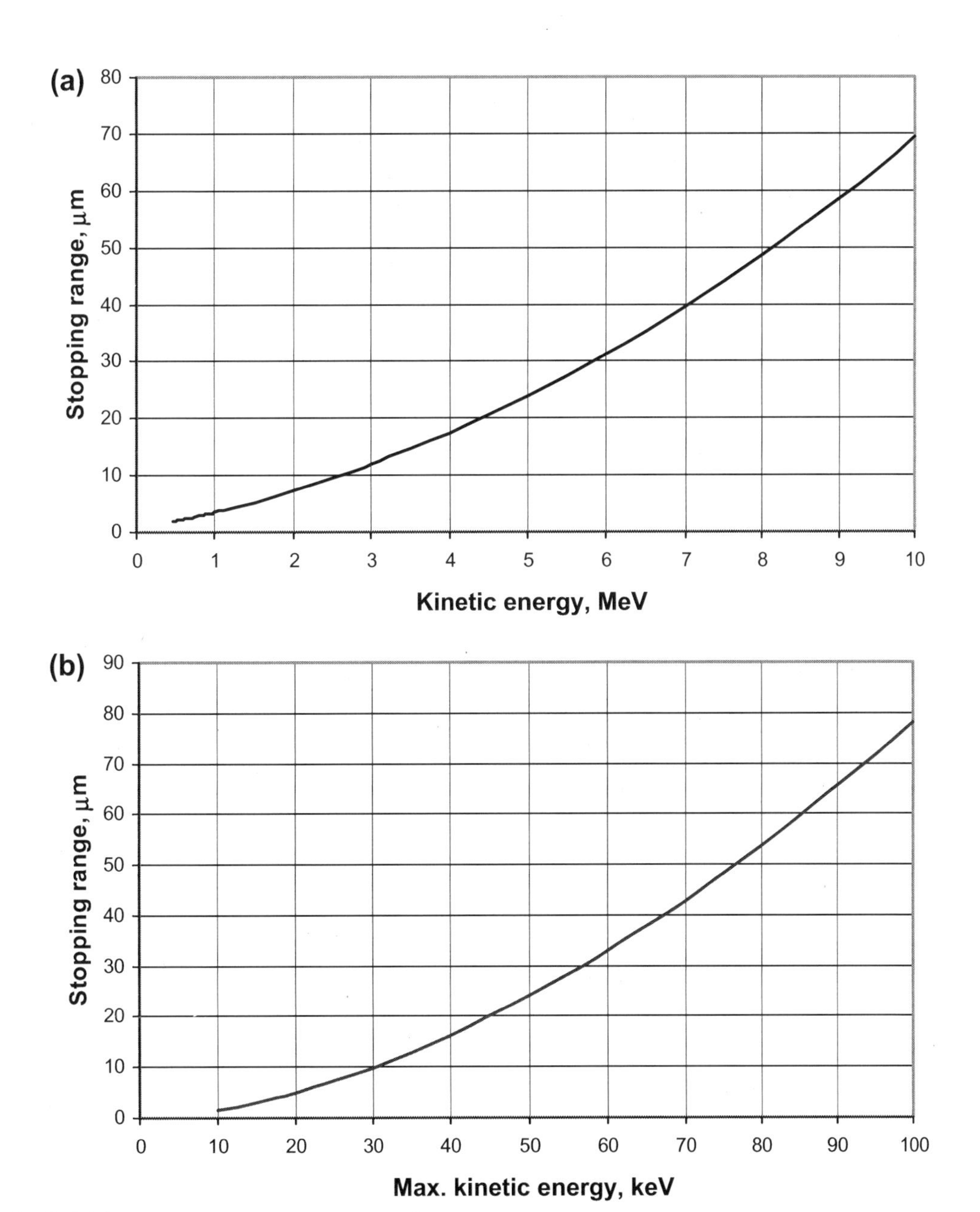

FIGURE 2.3.2

The stopping range in silicon for: (a) α-particles [5], and (b) β-particles [6]

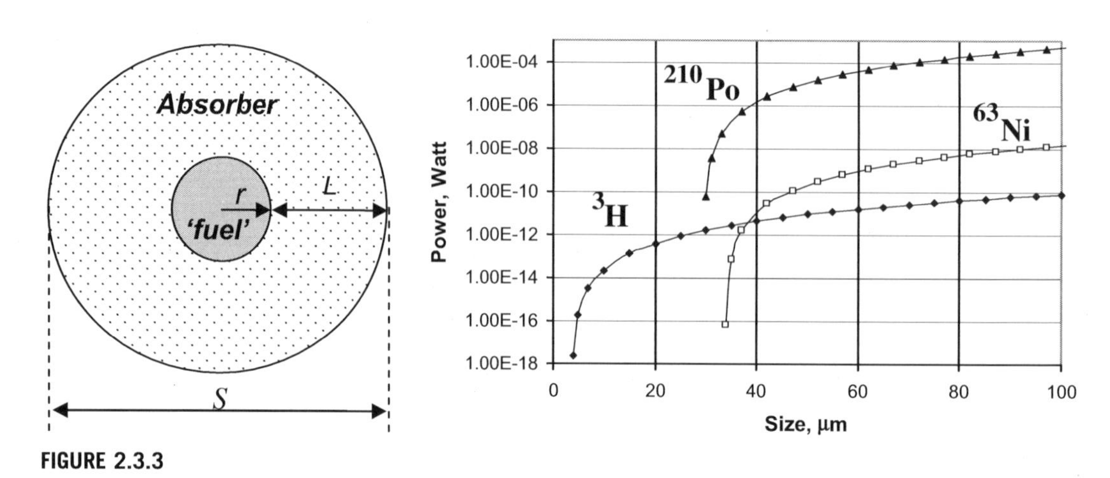

FIGURE 2.3.3

Size constraints of a radioisotope energy source: (a) a schematic drawing of the energy source, consisting of a 'fuel compartment' and an 'energy absorber', (b) maximum power as a function of total size

Assuming $T_c = 300$ K (ambient) and $T_h = 400$ K (a practical operational limit for Si VLSI), the Carnot efficiency $\eta = 25\%$.

Interaction of α- and β-particles with semiconductor-absorbing matter results in the creation of electron-hole pairs which can be separated at different electrodes and, thus, create a potential difference between the electrodes. This is analogous to the effect of incident photons in photovoltaic cells and is referred to as 'alphavoltaics' and 'betavoltaics'.

One of the factors determining the efficiency bound for betavoltaics is the electron-hole creation energy, $E_{e\text{-}h}$, which depends on the semiconductor bandgap E_g. The *minimum* $E_{e\text{-}h}$ (impact ionization threshold) is $E_{e\text{-}h\ min} \sim 3/2E_g$, resulting from the momentum conservation requirement. As was discussed in [43], the *average* energy of the electron-hole pair creation can be approximately written as $\overline{E}_{e-h} \sim 2E_{e-h_{\min}}$. Next, one needs to take into account that a fraction Δ of the particle kinetic energy goes directly to the lattice vibrations; thus:

$$\overline{E}_{e-h} \sim 3E_g + \Delta \qquad (2.3.9)^2$$

Relation (2.3.9) approximately holds for all semiconductors and for all types of incident particles (e.g. α, β, γ). For example, in silicon, the pair-creation energy is 3.6 eV for α-particles and 3.8 eV for electrons.

The maximum conversion efficiency can be estimated as:

$$\eta = \frac{E_g}{\overline{E}_{e-h}} = \frac{E_g}{3E_g + \Delta} \qquad (2.3.10a)$$

This equation can be re-written by dividing both the numerator and denominator by E_g:

$$\eta = \frac{1}{3 + \frac{\Delta}{E_g}} \qquad (2.3.10b)$$

[2]As it was shown in [9], factor of 14/5 in (2.3.9) gives a more accurate approximation, but for the expository level of the discussion in this study we use factor of 3.

Table 2.3.2 Recent results on alpha- and betavoltaic radioisotope energy sources

Semi-conductor	E_g, eV	Area	I_{sc}	V_{oc}	P	η	Source	Year, Ref.
SiC	3.2	1 cm^2	2 μA	2 V	~1 μW	4.5%	^{33}P	2006 [3]
SiC	3.2	0.5 mm^2	21 pA	~0.55 V	~5 pW	–	^{63}Ni	2007 [2]
InGaP	1.85	1 cm^2	0.5 nA	0.4 V	0.01 nW	0.04%	^{241}Am	2006 [9]

As follows from (2.3.10b), higher theoretical efficiency is expected for larger E_g and the maximum efficiency $\eta_{max} \sim 33\%$ (for silicon, $\eta \sim 29\%$).

A typical betavoltaic cell consists of a radioisotope β electron emitter integrated with a semiconductor *pn*-junction and operates in a manner similar to how a photovoltaic cell operates. The efficiency of such a cell will be less than given by (2.3.10) due to geometrical reasons: the radioisotope emits particles *isotropically* over 4π steradian, while only particles emitted in the direction of the *pn*-junction are absorbed for conversion. Therefore, only about one-quarter of all emitted particles will be available for conversion. To enhance particle collection, two-sided cells have been proposed. Overall, if planar *pn*-junctions are used, the particles emitted in the direction perpendicular to the surface of the semiconductor will not be collected. The number of these non-collected particles is $2\pi/4\pi = 1/2$, and thus the upper bound for the efficiency of planar apha/betavoltaic devices is given by (2.3.10) divided by two or ~15%. Of course this number will be further decreased by parasitic effects in the semiconductor (not considered here). In principle, the spherical device geometry of Figure 2.3.3 would considerably enhance the conversion efficiency; however, the feasibility of such a design is unclear.

2.3.3 Current status of miniature radioisotope energy sources

Table 2.3.2 presents several recent results on alpha- and betavoltaic radioisotope energy sources. The characteristic parameters of the sources are the open-circuit voltage V_{oc}, the short-circuit current I_{sc}, the conversion efficiency η, and the power *P*.

Recently, a MEMS radioisotope micropower generator was demonstrated [1,11–13]. The process is as follows. Emitted charged particles (e.g. electrons) are collected by a microcantilever. As a result, electrostatic force is produced between the cantilever and the radioisotope film. The cantilever is attracted toward the film, and thus produces mechanical movement, which could be further converted into electricity by, e.g., the piezoelectric effect [12]. The volumetric energy density of such a generator powered by ^{63}Ni isotope was estimated to be 5.0×10^{-2} J/cm^3[11] and thus, about 50 pJ of energy can be stored in a 10-μm cube.

2.4 REMARKS ON ENERGY HARVESTING

In the context of the nanomorphic system, energy harvesting refers to the collection of energy from external sources and its conversion into electrical form to power the system [48,69]. The biofuel cell, discussed in Section 2.2.4, is an example of harvesting from unintentional sources, i.e. from the ambient environment. Energy can also be harvested from intentional sources that transmit energy to the nano-morphic cell for conversion and conditioning. In the spirit of autonomous system operation, delivery by directed external energy sources covered herein does not include electrical connections, e.g. wires. The external energy accessible for harvesting can be in the form of radiation (solar, laser, RF), mechanical (vibrations), thermal, etc. As will be shown in this section, the amount of energy available for harvesting is fundamentally limited by the level of energy available in the 'safe' ambient environment, e.g. as defined by various regulatory agencies. Therefore, the remarks below are more concerned with the energy available for harvesting rather than details of operation of particular energy-harvesting devices.

2.4.1 Radiation

Solar energy

One sustainable source of energy in Earth's environment is the sun, whose energy can be converted into electricity by a photovoltaic cell. The operation of the photovoltaic cell is based on interaction of light with semiconductor-absorbing matter, creating electron-hole pairs that can be separated at different electrodes, and thus create a potential difference between the electrodes.

Operation of the photovoltaic cells and their use for energy production has been thoroughly studied [49,50]. Table 2.4.1 illustrates the solar power available at the Earth's surface under various environmental conditions. According to this table, the maximum energy available to a photovoltaic system is given by the product of the solar energy density and the area of the photovoltaic cell. The maximum solar irradiance is ~1000 W/m^2; thus, for the system size of ~10 μm we obtain:

$$P \sim 1000 \frac{W}{m^2} \cdot (10^{-5} m)^2 \approx 10^{-7}\, W$$

Table 2.4.1 Solar irradiance for typical terrestrial conditions

	T_{amb} (°C)*	P_{peak} (W/m^2)**	Note
High irradiance, high temperature (HIHT)	+45	1100	Very hot and sunny summer day in a desert location
Normal irradiance, cool environment (NICE)	+18	1000	Summer day in a cool coastal region
High irradiance, low temperature (HILT)	−1...+3	1000	Early spring day in Central Europe
Medium irradiance, high temperature (MIHT)	+27... +33	600	Very hot but cloudy and humid summer day
Medium irradiance, medium temperature (MIMT)	+7...+14	350	Warm and cloudy day
Low irradiance, low temperature (LILT)	−0.6...0	260	Winter day in Central Europe

*T_{amb}– ambient temperature.
**P_{peak}– peak irradiance.
(data taken from [51])

which is about an order of magnitude less than the power density attainable from a galvanic cell of the same size. Note that this estimate assumes 100% conversion efficiency, which is clearly overly optimistic. The overall efficiency of practical photovoltaic systems ranges between 11% and 30%, which would substantially reduce the output power.

Laser energy

In principle, one can think of powering an autonomous microsystem with a highly collimated laser beam. If the beam is focused to the spot diameter of ~10 μm, a beam power density of ~1 W/cm^2 would supply the power of ~1 μW to the nanomorphic cell, i.e. about the same power as a galvanic cell. Limits on laser power exposure by humans are subject to safety regulations. Figure 2.4.1 shows maximum permissible exposure limits for laser radiation of humans for different wavelengths and exposure times [51].

Of course direct laser powering is not an option for operations where the target location is not discernible; however, it might be considered for some applications. Photovoltaic laser power converters have been demonstrated for larger-scale systems [52, 53]. In [53], a near-infrared laser ($\lambda = 810$ nm) was used to recharge a lithium battery in a device implanted in a live rat. A photovoltaic cell array embedded under the skin received the near-infrared light and generated electricity sufficient to charge a battery. The total detection area of the photocells was 2.1 cm^2 and the power density of incoming radiation was 22 mW/cm^2 (limited by the permissible skin temperature rise). It was shown that 17 min of laser illumination can provide enough energy to run a commercial cardiac pacemaker

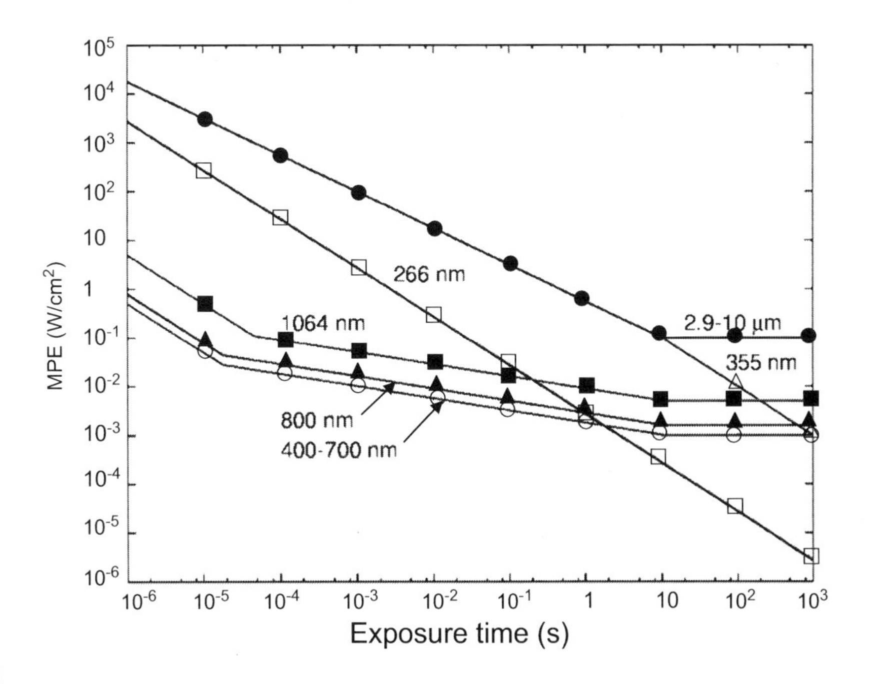

FIG. 2.4.1

Maximum permissible exposure for skin exposure to a laser beam [5]

for 24 h. It is important to note that in the above experiment [7], the dominant cause of the skin temperature rise was heating of the photocell array rather than the direct absorption of laser irradiation by the skin.

RF/microwave energy

The concept of wireless transmission of electrical energy via RF/microwave radiation is more than a century old, stemming back to Tesla's experiments. Today it is routinely used in radio-frequency identification (RFID) tags. Possibilities of using RF/microwave energy for powering small autonomous devices are also being explored [48,55].

The main issue associated with powering a microsystem with radiofrequency (RF) radiation is that when the radiation wavelength is larger than the receiving antenna size (which would be comparable to the 10-μm size of the nanomorphic cell), the absorption efficiency of the receiving antenna dramatically decreases. This problem is analyzed in detail in Chapter 5. For example, if the RF frequency is 300 GHz (wavelength of 1 mm), the efficiency of a 10 μm-long antenna is $\sim 10^{-4}$. Moreover, non-directional radiation from the source will also result in a dramatic increase in total required radiative energy as a function of distance between the transmitter and receiver (Frij's law discussed in Chapter 5). However, the levels of RF radiation in the human living environment are strictly regulated. As follows from the IEEE standards on the maximum permissible exposure limits for RF electromagnetic fields (Fig. 2.4.2), the maximum permissible exposure of radiation with wavelength of 1 mm is 10 mW/cm^2. Even if one assumes 100% receiving efficiency, the maximum power delivered to the microsystem is 10^{-8} W, which is less than that of a galvanic cell. With $\sim 10^{-4}$ antenna efficiency the power delivery is on the order of 10^{-12} W.

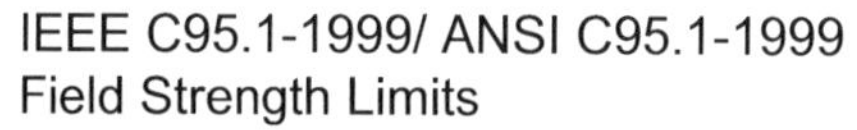

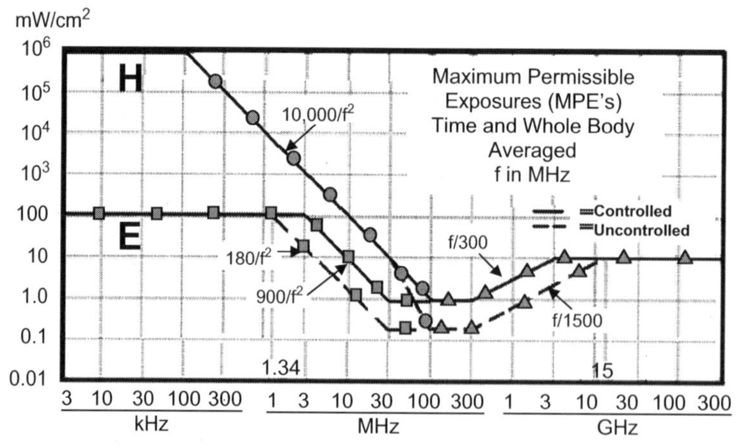

FIGURE 2.4.2

Maximum permissible exposure limits for RF electromagnetic fields [9]

Kinetic energy

The harvesting of kinetic energy, e.g. in the form of vibration, is sometimes regarded as an attractive source for powering wearable systems [48,57–62,69]. The ambient vibration is usually characterized by its amplitude x, frequency f, and acceleration magnitude a. Examples of typical vibration sources are shown in Table 2.4.2. The *exposure action value* (EAV) is a limit set on occupational exposure to vibration/noise where beyond those values, employers must take steps to monitor the exposure levels. An upper extreme limit *exposure limit value* (*ELV*) is provided as well to give a margin. Figure 2.4.3 depicts admissible exposure limits for vibration.

Below, a simple order-of-magnitude estimate of the vibrational energy available from the ambient environment is given. The kinetic energy of a body of mass m moving with velocity u is

$$E_k = \frac{mu^2}{2} \tag{2.4.1a}$$

The energy density (energy per unit volume) is

$$E_{K_V} = \frac{E_k}{v} = \frac{\rho u^2}{2} \tag{2.4.1b}$$

where ρ is the density of the body or the media. Vibration is represented by a periodic (not necessarily harmonic) movement with frequency f (period $\Theta = 1/f$), where the amplitude x is reached in $t = \Theta/4$ (quarter period). Thus, the characteristic velocity of the process is

$$u \sim \frac{x}{t} = \frac{4x}{\Theta} = 4xf \tag{2.4.2}$$

Similarly, the characteristic acceleration is

$$a \sim \frac{u}{t} = \frac{4u}{\Theta} = 4uf = 16xf^2 \tag{2.4.3}$$

From (2.4.1b), (2.4.2), and (2.4.3), the vibrational energy density is

$$E_k = 8\rho x^2 f^2 = \frac{\rho a^2}{32 f^2} \tag{2.4.4}$$

The power density of ambient vibration can be obtained by dividing (2.4.4) by $t = T/4 = 1/4f$:

$$P_V = \frac{E_{K_V}}{t} = 32\rho x^2 f^3 = \frac{\rho a^2}{8f} \tag{2.4.5}$$

For a numerical estimate, let $f \sim 10$ Hz and $\rho \sim 20$ g/cm^3, which is close to the densest natural solids, such as gold ($\rho_{Au} = 19300$ kg/m^3), platinum ($\rho_{Pt} = 21\,450$ kg/m^3), iridium ($\rho_{Ir} = 22\,500$ kg/m^3), and osmium

Table 2.4.2 Examples of typical vibration sources [64]

Source	Frequency, Hz	Amplitude, Cm
Air compressors	4–20	10^{-2}
Pumps	5–25	10^{-3}
Transformers	50–400	10^{-4}
Foot traffic	0.55–6	10^{-5}

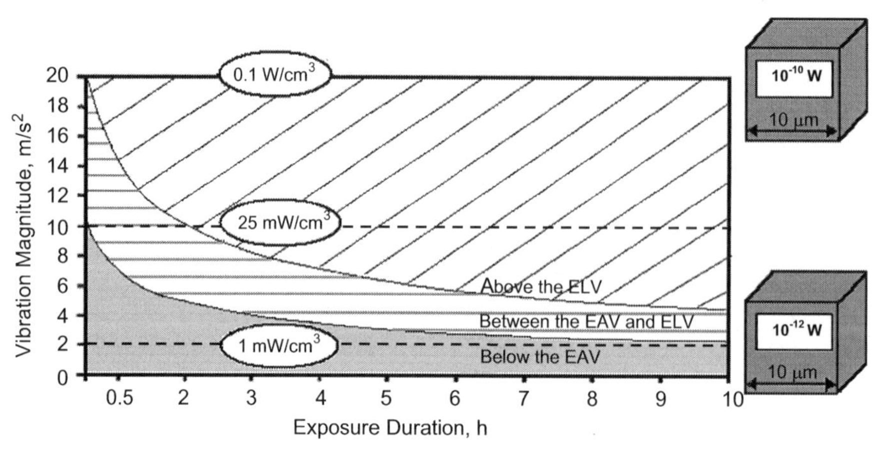

FIGURE 2.4.3

Admissible exposure limits for vibration

(adapted from [16])

($\rho_{Os} = 22\,610\,\text{kg/m}^3$). The numbers of characteristic vibrational power densities calculated using (2.4.5) are shown in Figure 2.4.3. It can be seen that for the vibration levels below the exposure action value (bottom zone), the vibration power available for harvesting is about 1 mW/cm^3. For a nanomorphic cell 10 μm in size, this would result in ~1 pW. Even for vibration levels above the exposure limit value (top zone), the available power is below 0.1 W/cm^3 or 100 pJ per nanomorphic cell.

The above analysis deals only with the kinetic (vibrational) energy available from the ambient environment, without any references to the efficiency of the energy-harvesting device (100% efficiency is assumed). In practical demonstrations, typical output power of the vibration energy generators is about 100 μW/cm^3[10–15], which is about one order of magnitude lower than the upper bound estimates obtained in this section.

Thermal energy

If a temperature difference exists within a material structure, thermal energy can be extracted, e.g. by the thermoelectric effect. Miniature thermoelectric generators are currently being explored as an autonomous power source for some wearable electronic devices [48,65–69]. The principle of thermoelectricity is illustrated in Figure 2.4.4 (see also Chapter 4, Box 4.5): electrons from the hot side have larger kinetic energy and therefore move towards the cold side. As a result, excess negative charge is accumulated on the cold side and therefore an electrical voltage is created between the hot and cold sides. If now the two sides are electrically connected to an external load, power will be delivered to the load (Fig. 2.4.5).

The maximum electric power (upper bound) P_{max} which can be extracted from an input heat flux Q^+ is given by the Carnot efficiency already discussed in Section 2.3:

$$P_{\max} = Q^+\left(\frac{T_h - T_c}{T_h}\right) = Q^+\frac{\Delta T}{T_h} \tag{2.4.6a}$$

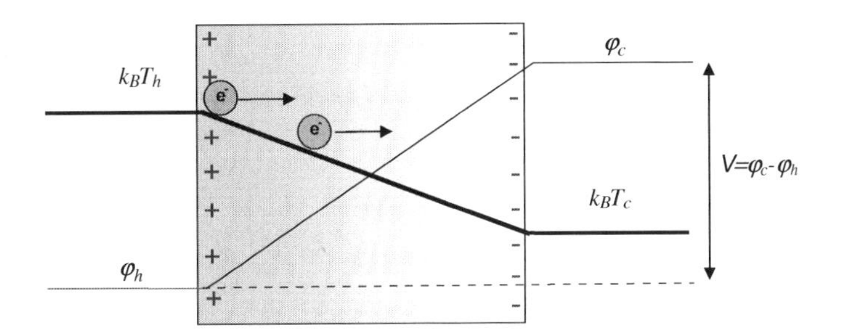

FIGURE 2.4.4

Thermoelectric effect: Difference in temperatures between hot (T_h) and cold (T_c) side of a material structure results in electrical potential difference $V = \varphi_c$-φ_h between the two sides

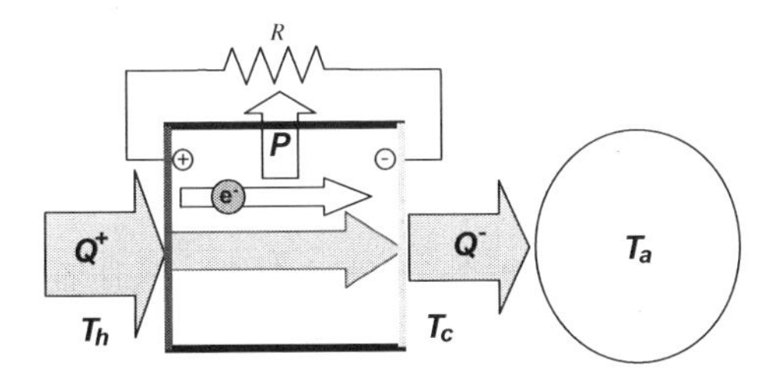

FIGURE 2.4.5

Thermoelectric generator

The 'unused' heat energy constitutes the output heat flux Q^-, which must be removed from the system:

$$Q^- = Q^+ - P_{\max} = Q^+ \frac{T_c}{T_h} \tag{2.4.6b}$$

Thus, the maximum electrical power delivery by a thermoelectric generator is limited by the amount of heat passing through the structure. Note that the output heat flux Q^- is fundamentally limited by the properties of the ambient heat sink, which in terrestrial conditions is usually air or water at $T_a = 300$ K. Therefore it is instructive to express the maximum electric power P_{max} as a function of the output heat flux Q^-:

$$P_{\max} = Q^- \frac{T_h - T_c}{T_c} \tag{2.4.6c}$$

It is clear from (2.4.6c) that the performance of a thermoelectric generator is limited by the heat removal rate from the system. In this regard, it is important to note that while for larger-scale thermoelectric converters, special techniques are often utilized to significantly improve the heat flow, such as heat spreaders, liquid heat exchangers, radiators, etc. [67]. These techniques serve as an agent between the system and the ambient environment and can occupy significant volume. In the case of the nanomorphic cell, where space is strictly constrained, volume extensions would utilize valuable space needed for other functionalities. Thus, the nanomorphic thermoelectric generator should function as a completely passive energy-harvesting system with direct heat exchange with the surrounding, typically air or water at T_a ~ 300 K. Furthermore, for many applications, especially with living systems, direct contact with an ambient water reservoir is not feasible, leaving air as the remaining option. On the other hand, for exterior application, conditions simulating forced air convection may exist, such as wind chill. Another limitation for nanomorphic thermoelectric generators is that relatively small temperature gradients are likely, especially for applications with living systems (for example, utilizing human warmth). In view of the above, the limits of heat removal will be estimated for the following assumptions:

1. Main mechanism for heat transfer: Forced convection to ambient air
2. Ambient temperature: $T_a = 300$ K
3. Max. temperature difference: $\Delta T = T_h - T_a \sim 5$ K.

To estimate limits of $Q^{\cdot}$, we consider an ideal case of heat transfer to the ambient environment, which represents an abstraction for forced convection [70, 71]. The heat removal is mainly due to the transfer of thermal (kinetic) energy by the collision of the 'hot' atoms/molecules of the body surface (at $T = T_c$) with 'cold' ambient molecules (air or water) at $T_a = 300$ K. After collision with a 'hot' surface atom, the ambient molecule absorbs additions energy ΔE and therefore becomes 'warm', i.e. $T > 300$ K. These 'warm' ambient molecules will reduce the energy transfer rate (in practice this corresponds to the local increase in the ambient temperature resulting in a decrease of cooling efficiency). To deal with this, forced cooling needs to be employed, where each 'warm' molecule is forcefully replaced by a fresh 'cold' ambient molecule at $T_a = 300$ K. In this case, the cooling rate is determined by the supply rate of 'cold' ambient molecules.

To estimate the maximum cooling rate by active gas/liquid cooling, we consider a chain of atoms at temperature T_h and a flow of 'cold' ambient molecules falling on the surface (Fig. 2.4.6). At the surface, the 'cold' ambient molecules of mass m_a collide with the 'hot' atoms of mass m_h, collect some of the heat energy, and move away from the surface atoms. 'Cold' ambient molecules in the incoming flow would have a velocity component, *u,* normal to the surface. The normal velocity is made up of two parts: the *directed flow* component u_0 and *thermal* component u' (kinetic energy ~ $k_B T_a = \frac{1}{2} m_a u'^2$):

$$u' = \sqrt{\frac{2k_B T_a}{m_a}} \tag{2.4.7a}$$

$$u = u_0 \pm u' = u_0 \pm \sqrt{\frac{2k_B T_a}{m_a}} \tag{2.4.7b}$$

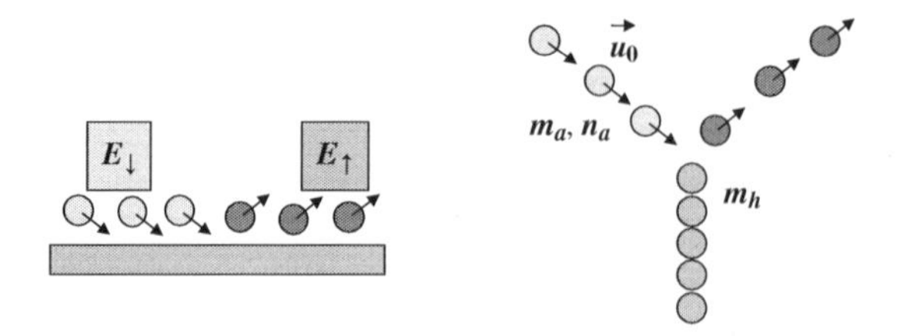

FIGURE 2.4.6

A depiction of the heat transfer from a heated surface by interaction with ambient molecules

where m_a is the molecular mass of the ambient molecules and T_a is the ambient temperature. Since the direction of the thermal component is random, the average velocity of incoming ambient molecules is $\langle u \rangle = u_0$.

After colliding with the surface, the ambient molecules absorb certain energy $\langle E \rangle$from the surface atoms. The corresponding heat transfer rate (heat power) per surface atom is

$$p = \frac{\langle \Delta E \rangle}{\Delta t} \tag{2.4.8}$$

For an estimate of $\langle E \rangle$, consider the set of possible elastic collisions of two bodies with different masses m_1 and m_2 (see Appendix). When considering a collision of two atoms with velocities v and $u = u_0 \pm u'$ we have eight possible realizations, as shown in Figure 2.4.7.

Since all realizations have equal probability, the expected energy transfer is the mean value:

$$\langle \Delta E \rangle = \frac{\Delta E_{\uparrow\downarrow+} + \Delta E_{\uparrow\downarrow-} + \Delta E_{\uparrow\uparrow+} + \Delta E_{\uparrow\uparrow-} + 0 + 0 + 0 + 0}{8} \tag{2.4.9}$$

Expressions for each term in (2.4.9) are given by (A4) and are displayed in Fig. 2.4.7. Using them and taking into account (2.4.9) gives:

$$\langle \Delta E \rangle = \mu(2k_B \Delta T - m_a u_0^2) \tag{2.4.10}$$

where $\mu = \frac{m_a m_h}{(m_a + m_h)^2}$ is the reduced mass and $\Delta T = T_h - T_a$.

The time interval between two collisions of the ambient molecules with surface atoms is

$$\Delta t = \frac{l}{\langle u \rangle} = \frac{n_a^{-\frac{1}{3}}}{u_0} \tag{2.4.11}$$

where n_a is the atomic density of the ambient media (e.g. air or water).

From (2.2.10) and (2.4.11), the heat transfer rate per channel is

$$q = \frac{\langle \Delta E \rangle}{\Delta t} = \mu(2k_B \Delta T - m_a u_0^2) \cdot u_0 \cdot n_a^{\frac{1}{3}} \tag{2.4.12}$$

and per unit area is

$$Q = q \cdot n_a^{\frac{2}{3}} = \mu(2k_B \Delta T - m_a u_0^2) \cdot u_0 \cdot n_a \tag{2.4.13}$$

The maximum of (2.4.13) is reached when

$$\frac{dQ}{du} = 0 \tag{2.4.14}$$

from which there results:

$$u_{0\max} = \sqrt{\frac{2k_B \Delta T}{3m_a}} \tag{2.4.15a}$$

and

$$Q_{\max} = \frac{4\mu n_a}{3} \sqrt{\frac{2(k_B \Delta T)^3}{3m_a}} \tag{2.4.15b}$$

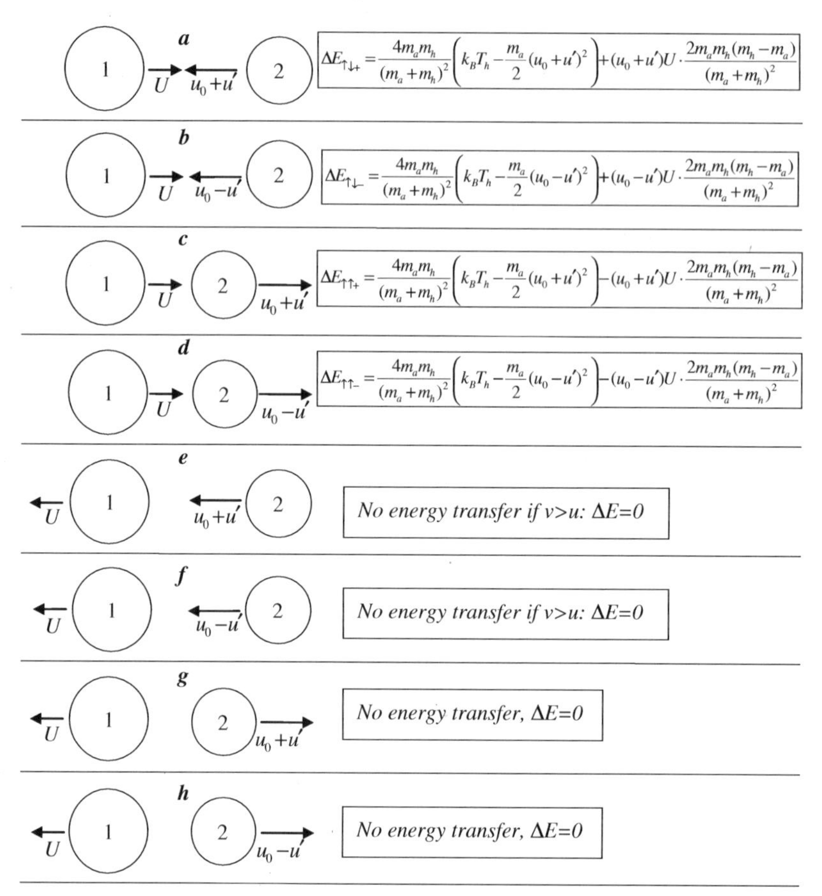

FIGURE 2.4.7

Energy transfer between two elastically colliding balls

A numerical estimate of Q_{max} can be made for a silicon surface ($m_h = 28$ a.m.u.) and air ($m_a = 29$ a.m.u., $n_a = 2.7 \times 10^{19}$ cm^{-3}). The temperature of the 'hot' surface is $T_h = 305$ K and the temperature of ambient air is $T_a = 300$ K. For these conditions, the plot $Q(u_0)$is shown in Figure 2.4.8 and

$$Q_{\max} = 1.92\frac{W}{cm^2}$$

Thus, the maximum electric power delivery of a 10 μm-sized thermoelectric generator, calculated from (2.4.6c) for $T_h = 305$ K and $T_c = T_a = 300$ K, is

$$P_{\max} \sim 10^{-8} W$$

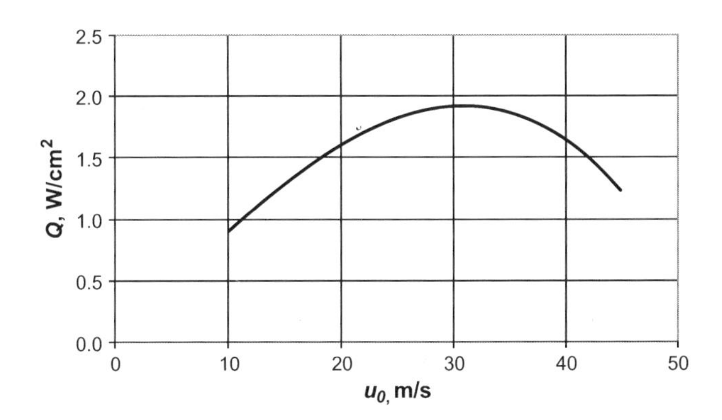

FIGURE 2.4.8

Maximum heat removal rate for forced air cooling ($T_a = 300$ K) of a Si surface ($T_h = 305$ K)

2.5 SUMMARY

In this chapter upper bounds for energy and power of nanomorphic implementation of different energy sources were derived based on fundamental physics of operation and assuming ideal conditions. Results of the derivations in this chapter are summarized in Table 2.5.1. It can be seen that the electrochemical galvanic cell provides the best combination of stored energy and power delivery among the options considered. In subsequent chapters the galvanic cell power and energy estimates will be used as a reference for available energy and power.

Table 2.5.1 Energy and power of nanomorphic implementation of different energy sources

	Stored energy, J	Power, W
Galvanic cell	10^{-5}	10^{-6}
Supercapacitor	10^{-7}	1
Radioisotopes	10^{-5}	10^{-14}
(Bio) Fuel cell	Sustainable	10^{-8}
Solar	Sustainable	10^{-7}
Laser	Sustainable	10^{-7}
RF	Sustainable	$10^{-8}/10^{-12}$
Vibration	Sustainable	10^{-12}
Thermal	Sustainable	10^{-8}

APPENDIX: A KINETIC MODEL TO ASSESS THE LIMITS OF HEAT REMOVAL

A simple approach to represent heat transfer in a body is to consider the transfer of energy that occurs when two masses collide. In Figure A1, energy is transferred from moving Ball 1 with mass m_1 to stationary Ball 2 with mass m_2.

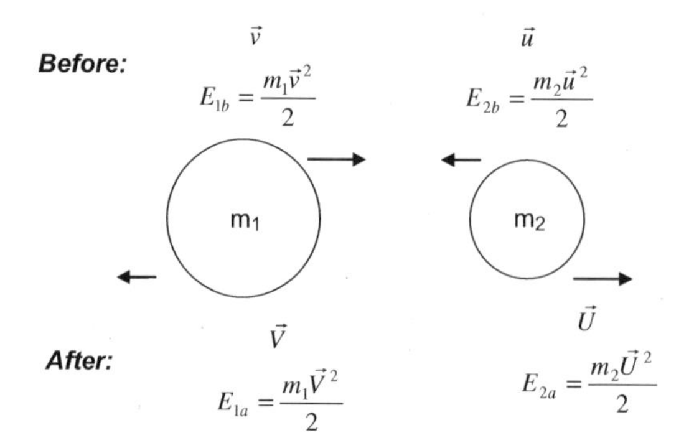

FIGURE A1

Two colliding balls of arbitrary masses and initial velocities

The energy transfer in this system can be calculated from the momentum and energy conservation:

$$m_1\vec{v} + m_2\vec{u} = m_1\vec{V} + m_2\vec{U}\; \frac{m_1\vec{v}^2}{2} + \frac{m_2\vec{u}^2}{2} = \frac{m_1\vec{V}^2}{2} + \frac{m_2\vec{U}^2}{2} \tag{A1}$$

where v and u are the corresponding velocities of Balls 1 and 2 before collision, while V and U are the corresponding velocities after collision.

Solution of (A1) for V and U is:

$$\begin{aligned} V &= v\frac{m_1 - m_2}{m_1 + m_2} + u\frac{2m_2}{m_1 + m_2} \\ U &= v\frac{2m_1}{m_1 + m_2} + u\frac{m_2 - m_1}{m_2 + m_1} \end{aligned} \tag{A2}$$

Let the energy of Ball 1 before the collision be E_{1b} and after the collision E_{1a}. Similarly, the energy of Ball 2 before and after collision is correspondingly E_{2b} and E_{2a}.

Energy change in Ball 1 as a result of the collision is:

$$\begin{aligned} \Delta E_1 &= E_{1b} - E_{1a} = \frac{m_1}{2}(v^2 - V^2) = \frac{m_1}{2}\left(v^2\frac{4m_1m_2}{(m_1+m_2)^2} - u^2\frac{4m_2^2}{(m_1+m_2)^2} - uv\frac{4m_2(m_1-m_2)}{(m_1+m_2)^2}\right) \\ &= \frac{4m_1m_2}{(m_1+m_2)^2}\left(v^2\frac{m_1}{2} - u^2\frac{m_2}{2}\right) - uv\frac{2m_1m_2(m_1-m_2)}{(m_1+m_2)^2} \end{aligned} \tag{A3}$$

or

$$\Delta E = \frac{4m_1m_2}{(m_1+m_2)^2}(E_{1b} - E_{2b}) - uv\frac{2m_1m_2(m_1-m_2)}{(m_1+m_2)^2} \tag{A4}$$

LIST OF SYMBOLS

Symbol	Meaning
a	Acceleration
A	Surface area
a_s	Specific surface area
b	Constant
c_m	Gravimetric specific capacitance
c_s	Specific capacitance
C	Capacitance
C_f	Faradic capacitance
d	Thickness
e	Electron charge, $e = 1.6 \times 10^{-19}$ C
E	Energy
E_g	Semiconductor bandgap
f	Frequency
I	Current
k	Constant
k_B	Boltzmann constant, $k_B = 1.38 \times 10^{-23}$ J/K
K	Dielectric constant
l	Length
$l_{a\text{-}a}$	Nearest-neighbor distance between atoms
L	Stopping range
m	Mass
M	Molar mass
n, n_{at}, n_{3D}	Atomic density
n_{1D}	Number of atoms per unit of length
n_s, n_{2D}	Surface concentration of atoms
n_m	Number of atoms per unit mass
N_A	Avogadro's Number, $N_A = 6.022 \times 10^{23}$ mol^{-1}
N_{at}, $N(t)$, N_0	Number of atoms
N_{el}	Number of electrons
N_{SW}	Number of switching events
p	Heat transfer rate per atom
P	Power
q	Electric charge
Q	Heat flux
r	Radius
R	Resistance
S	Size

(*Continued*)

Symbol	Meaning
t	Time
$t_{1/2}$	Radionuclide half life
T	Absolute temperature
v	Volume
V	Voltage
w	Width
x	Amplitude
δ	Distance between the neighboring atomic planes (for graphite $\delta = 0.335$ nm)
Δ	Particle kinetic energy transferred to lattice vibrations
ε	Energy released per atom
ϵ_0	Permittivity of vacuum, $\varepsilon_0 = 8.85 \times 10^{-14}$ F/cm
η	Efficiency
Θ	Period of oscillation
μ	Reduced mass
ρ	Density of substance
σ	Conductivity
τ	Radionuclide mean life time
φ	Electric potential
~	Indicates order of magnitude

References

[1] S. Tajima, Aluminum and manganese as anodes for dry and reserve batteries, J. Power Sources 11 (1984) 155–161.

[2] Q. Li, N.J. Bjerrum, Aluminum as anode for energy storage and conversion: a review, J. Power Sources 110 (2002) 1–10.

[3] W.S.D. Wilcock, P.C. Kauffman, Development of a seawater battery for deep-water applications, J. Power Sources 66 (1997) 71–75.

[4] M.A. Klochko, E.J. Casey, On the possible use of titanium and its alloys and compounds as active materials in batteries: A review, J. Power Sources 2 (1977/78) 201–232

[5] S. Downey, Small Energy Sources, SRC/NSF Forum on Nano-Morphic Systems: Processes, Devices, and Architectures, Stanford University, Stanford, CA, November 8–9, 2007.

[6] A. Heller, Potentially implantable miniature batteries, Anal. Bioanal. Chem. 385 (2006) 469–473.

[7] W. Shin, J. Lee, Y. Kim, H. Steinfink, A. Heller, Ionic conduction in $Zn_3(PO_4)_2$-$4H_2O$ enables efficient discharge of the zinc anode in serum, J. Amer. Chem. Soc. 127 (2005) 14590.

[8] W. Shin, Miniature bio-fuel cell and Zn-Ag/AgCl battery in physiological condition, SRC/NSF Forum on Nano-Morphic Systems: Processes, Devices, and Architectures, Stanford University, Stanford, CA, November 8–9, 2007.

[9] N. Mano, F. Mao, W. Shin, T. Chen, A. Heller, A miniature biofuel cell operating at 0.78 V, Chem. Commun. (2003) 518–519.

[10] A. Heller, Miniature biofuel cells, Phys. Chem. Chem. Phys. (2004) 209–216.

[11] N. Mano, F. Mao, A. Heller, Characteristics of a miniature compartment-less Glucose-O_2 biofuel cell and its operation in a living plant, J. Amer. Chem. Soc. 125 (2003) 6588–6594.
[12] D. Wu, R. Tucker, H. Hess, Caged ATP – Fuel for Bionanodevices, IEEE Trans. Adv. Pack. 28 (2005) 594.
[13] H. Hess, Biomolecular motors for directed assembly and hybrid devices, SRC/NSF Forum on Nano-Morphic Systems: Processes, Devices, and Architectures, Stanford University, Stanford, CA, November 8–9, 2007.
[14] Food and Nutrition Board of the Institute of Medicine, www.iom.edu
[15] E. Rojas, L.A. Herrera, L.A. Poirier, P. Ostrosky-Wegman, Are metals dietary carcinogens? Mutation Res. 443 (1999) 157–181.
[16] U.S. EPA Drinking Water Regulations, www.epa.gov
[17] R. Kötz, M. Carlen, Principles and applications of electrochemical capacitors, Electrochem. Acta 45 (2000) 2483–2498.
[18] A. Lewandowski, M. Galinski, Practical and theoretical limits for electrochemical double-layer capacitors, J. Power Syst. 173 (2007) 822–828.
[19] E. Frackowiak, Supercapacitors based on carbon materials and ionic liquids, J. Braz. Chem. Soc. 17 (2006) 1074–1082.
[20] G. Yushin, Integrated supercapacitors for nano-morphic systems, SRC/NSF Forum on Nano-Morphic Systems: Processes, Devices, and Architectures, Stanford University, Stanford, CA, November 8–9, 2007.
[21] J. Chmiola, G. Yushin, Y. Gogotsi, C. Portet, P. Simon, P.L. Taberna, Anomalous increase in carbon capacitance at pore sizes less than 1 nanometer, Science 313 (2006) 1760–1763.
[22] K.A. Williams, P.C. Eklund, Monte Carlo simulations of H_2 physisorption of finite-diameter carbon nanotube ropes, Chem. Phys. Lett. 320 (2000) 352–358.
[23] M. Winter, R.J. Brodd, What are batteries, fuel cells, and supercapacitors? Chem. Rev. 104 (2004) 4245–4269.
[24] L. S-Kou, L. N-Wu, Investigation of pseudocapacitive charge-storage reaction of $MnO_2 \cdot nH_2O$ supercapacitors in aqueous electrolytes, J. Electrochem. Soc. 153 (2006) A1317–1324.
[25] L. S-Kou, L. N-Wu, Electrochemical capacitor of $MnFe_2O_4$ with NaCl electrolyte, Electrochem. and Solid-State Lett. 8 (2005) A495–A499.
[26] N.A. Choudhury, S. Sampath, A.K. Shukla, Gelatin hydrogel electrolytes and their application to electrochemical supercapacitors, J. Electrochem. Soc. 155 (2008) A74–A81.
[27] H. J-Sung, S.-J. Kim, K.-H. Lee, Fabrication of microcapacitors using conducting polymer microelectrodes, J. Power Sources 124 (2003) 343–350.
[28] H. J-Sung, S.-J. Kim, S.-H. Jeong, E.-H. Kim, K.-H. Lee, Flexible micro-supercapacitors, J. Power Sources 162 (2006) 1467–1470.
[29] P.M. Raj, D. Balaraman, V. Govind, I.R. Abothu, L. Wan, R. Gerhardt, et al., Processing and dielectric properties of nanocomposite thin film supercapacitors for high-frequency embedded decoupling, IEEE Trans. Comp. Packaging Techn. 30 (2007) 569–578.
[30] L. S-Kuo, L. N-Wu, Composite supercapacitor containing tin oxide and electroplated ruthenium oxide, Electrochem. and Solid-State Lett. 6 (2003) A85–A87.
[31] H. Jimbo, N. Miki, Gastric-fluid-utilizing micro-battery for micro medical devices, Sensors and Actuators B134 (2008) 219–224.
[32] M. Sun, G.A. Justin, P.A. Roche, J. Zhao, B.L. Wessel, Y. Zhang, R.J. Sclabassi, Passing data and supplying power to neural implants, IEEE Eng, Medicine and Biology 25 (2006) 39–46.
[33] S. Kerzenmacher, J. Ducree, R. Zengerle, F. von. Stetten, Energy harvesting by implantable abiotically catalyzed glucose fuel cells, J. Power Sources 182 (2008) 1–17.
[34] E. Kjeang, N. Djilali, D. Sinton, Microfluidic fuel cells: A review, J. Power Sources 186 (2009) 353–369.
[35] A. Lal, Radioisotope Energy Sources, SRC/NSF Forum on Nano-Morphic Systems: Processes, Devices, and Architectures, Stanford University, Stanford, CA, November 8–9, 2007.

[36] M.V.S. Chandrashekhar, R. Duggirala, M.G. Spencer, A. Lal, 4H SiC betavoltaic powered temperature transducer, Appl. Phys. Lett. 91 (2007) 053511.

[37] C.J. Eiting, V. Krishnamoorthy, S. Rodgers, T. George, J.D. Robertson, J. Brockman, Demonstration of a radiation resistant, high efficiency SiC betavoltaic, Appl. Phys. Lett. 88 (2006) 064101.

[38] F.N. Huffman, J.C. Norman, Nuclear-fueled cardiac pacemakers, Chest 65 (1974) 667–672.

[39] J. Magill, Nuclides.net: An Integrated Environment for Computations of Radionuclides and their Radiation (Springer 2002).

[40] ASTAR: Stopping Power and Range Tables for Helium Ions, <http://physics.nist.gov/PhysRefData/Star/Text/ASTAR.html>

[41] ESTAR: Stopping Power and Range Tables for Electrons, <http://physics.nist.gov/PhysRefData/Star/Text/ESTAR.html>

[42] V.M. Balebanov, S.S. Moiseev, V.I. Karas', I.V. Karas', S.I. Kononenko, V.I. Kolesnik, Secondary-emission radioisotopic current source, Atomic Energy 84 (1998) 324–328.

[43] C.A. Klein, Bandgap dependence and related features of radiation ionization energies in semiconductors, J. Appl. Phys. 39 (1968) 2029–2038.

[44] C.D. Cress, B.J. Landi, R.P. Raffaelle, InGaP alpha voltaic batteries: Synthesis, modeling, and radiation tolerance, J. Appl. Phys. 100 (2006) 114519.

[45] H. Li, A. Lal, Self-reciprocating radioisotope-powered cantilever, J. Appl. Phys. 92 (2002) 1122–1127.

[46] A. Lal, R. Duggirala, H. Li, Pervasive power: A radioisotope-powered piezoelectric generator, IEEE Trans. Pervasive Computing (2005) 53–61.

[47] D. Yan, A. Lal, Silicon-on-insulator cantilevers as charge collectors for radioisotope micropower sources: design, fabrication and characterization, J. Micromech. Microeng. 16 (2006) 2100–2108.

[48] N.S. Hudak, G.G. Amatucci, Small-scale energy harvesting through thermolelectric, vibration, and radiofrequency power conversion, J. Appl. Phys. 103 (2008) 101301–101324.

[49] C.E. Backus (Ed.), Solar Cells, IEEE Press, New York, 1976.

[50] J. Nelson (Ed.), The Physics of Solar Cells, Imperial College Press, London, 2003.

[51] T. Zdanowicz, T. Rodziewicz, M. Zabkowska-Waclawak, Theoretical analysis of the optimum energy band gap of semiconductors for fabrication of solar cells for application in higher latitudes locations, Solar Energy Materials & Solar Cells 87 (2005) 757–769.

[52] Based on IEC 60825 standard. International Electrotechnical Commission. 2007. http://en.wikipedia.org/wiki/Laser_safety.

[53] J. Schubert, E. Oliva, F. Dimroth, W. Guter, R. Loeckenhoff, A.W. Bett, High-voltage GaAs photovoltaic laser power converters, IEEE. Trans. Electron Dev. 56 (2009) 170–175.

[54] K. Goto, T. Nakagawa, O. Nakamura, S. Kawata, An implantable power supply with an optically rechargeable lithium battery, IEEE Trans. Biomed. Eng. 48 (2001) 830–833.

[55] M. Mickle, M. Mi, L. Mats, C. Capelli, H. Swift, Powering autonomous cubic-millimeter devices, IEEE Antennas and Propagation Mag. 48 (2006) 11–21.

[56] http://www.rfsafetysolutions.com/IEEE_standard.htm

[57] P.D. Mitcheson, E.M. Yeatman, C.K. Rao, A.S. Holmes, T.C. Green, Energy harvesting from human and machine motion for wireless electronic devices, Proc. IEEE 96 (2008) 1457–1486.

[58] S. Roundy, P.K. Wright, J.M. Rabaey (Eds.), Energy Scavenging for Wireless Sensor Networks, Kluwer Academic Publishers, 2004.

[59] S.P. Beeby, M.J. Tudor, N.M. White, Energy harvesting vibration sources for microsystems applications, Meas. Sci. Technol. 17 (2006) R175–R195.

[60] B. Op het Veld, D. Hohlfeld, Valer Pop, Harvesting mechanical energy for ambient intelligent devices, Inf. Syst. Front. 11 (2009) 7–18.

[61] F. Cottone, H. Vocca, L. Gammaitoni, Nonlinear energy harvesting, Phys. Rev. Lett. 102 (2009) 080601.

[62] B. Marinkovic, H. Koser, Smart sand – a wide bandwidth vibration harvesting platform, A ppl. Phys. Lett. 94 (2009) 103505.

[63] Health and Safety Executive UK, HSEUK, http://www.hse.gov.uk/pubns/indg175.pdf

[64] Sources of Vibration - www.mellesgriot.com/pdf/CatalogX/X_31_3-7.pdf

[65] I. Boniche, B.C. Morgan, P.J. Taylor, C.D. Meyer, D.P. Arnold, Process development and material characterization of polycrystalline Bi_2Te_3, PbTe, and PbSnSeTe thin films on silicon for millimerter-scale thermoelectric generators, J. Vac. Sci. Technol. A26 (2008) 739–744.

[66] V. Leonov, T. Torfs, P. Fiorini, C. Van Hoof, Thermoelectric converters of human warmth for self-powered wireless sensor nodes, IEEE Sensors J. 7 (2007) 650–657.

[67] H.A. Sodano, G.E. Simmers, R. Dereux, D.J. Inman, Recharging batteries using energy harvested from thermal gradients, J. Intelligent Mat. Syst. And Struct. 18 (2007) 3–10.

[68] S. Dalola, M. Ferrari, V. Ferrari, M. Guizzetti, D. Marioli, A. Taroni, Characterization of thermoelectric modules for powering autonomous sensors, IEEE Trans. Instrum. Measur. 58 (2009) 99–107.

[69] J.A. Paradiso, T. Starner, Energy scavenging for mobile and wireless electronics, IEEE Pervasive Comp. 4 (2005) 18–27.

[70] A. Avila, R.K. Cavin, V.V. Zhirnov, H.H. Hosack, Fundamental limits of heat transfer, in: S.V. Garimela, A.S. Fleischer (Eds.), THERMES 2007: Thermal Challenges in Next Geneartion Electronic Systems, Millpress, Rotterdam, 2007.

[71] R.K. Cavin, V.V. Zhirnov, D.J.C. Herr, A. Avila, J. Hutchby, Research directions and challenges in nanoelectronics, J. Nanoparticle Res. 8 (2006) 841–858.

CHAPTER

Nanomorphic electronics

3

CHAPTER OUTLINE

3.1 INTRODUCTION

The autonomous microsystem must have an electronic control unit, which is assumed to be a specialized micro-scale computer. The capability of the unit is determined by its complexity (e.g. the device count) and its energy of operation. Moreover, as will be discussed in Chapter 5, the system's 'intelligence' (which is derived from a composition of logic and memory elements) needs to be maximized to reduce the communication costs. Thus system scaling limits need to be studied to understand the amount of 'intelligence' that could be expected from a volume of matter 10 μm in size.

In this chapter, fundamental scaling limits for micro-scale electronic circuits, constructed from binary devices and interconnects are investigated. These device and interconnect models are developed from basic physics using a generic energy barrier representation for binary devices. In order to estimate circuit performance, it is necessary to relate the device and interconnect system properties such as switching energy and switching times, number of electrons, etc., to the physical layout of the processor. It will be argued that the layout geometry can, in the limit, be viewed as an assembly of small *square (cubic) tiles*, to each of which is associated size, energy and travel (i.e. delay) time parameters derived from basic physics. The estimates of limits are performed for the device and interconnect systems assumed to operate at the threshold of failure (which corresponds to the lowest energy consumption). Implications for higher levels of device and system reliability are also considered.

BOX 3.1 MAIN CONCEPT: BINARY SWITCH

Information of an arbitrary kind and amount (such as letters, numbers, colors, or graphics specific sequences and patterns) can be represented by combination of just two states. The two states (known as binary states) are usually marked as digits **0** and **1**, thus information is represented in *digital* form.

A typical example of a binary switch is an *electrical switch* (Fig. B3.1). The switch can be only in two positions: open or closed (Fig. B3.1a). Connected binary switches can also perform logic operations such as AND (Fig. B3.1b), OR (Fig. B3.1c), etc. Certain combinations of logic operations are equivalent to arithmetic operations.

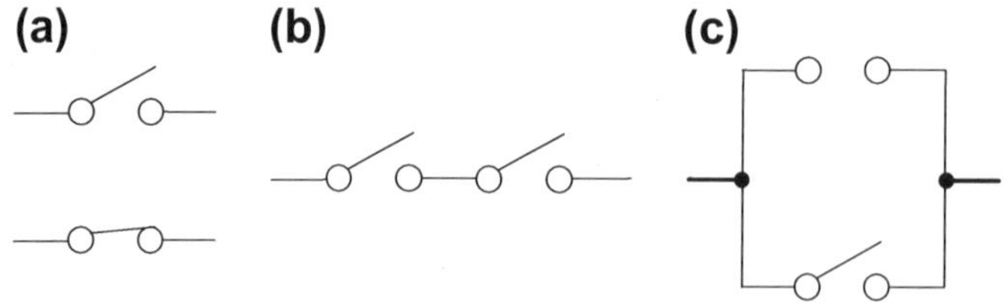

(a) (b) (c)

FIG. B3.1

Two states of a switch (a) and examples of implementing fundamental logic operations:(b) AND, and (c) OR

Modern integrated circuits (IC), also known as 'chips', contain millions and billions of controllable electronic switches, called transistors, which will be discussed later in this chapter (see Box 3.4).

3.2 INFORMATION AND INFORMATION PROCESSING

Information can be defined as a technically quantitative measure of the distinguishability of a physical subsystem from its environment [1,2]. One way to create distinquishable states is by the *presence or absence* of material particles (information carrier) in a given location. For example, information is encoded in DNA through specific locations of certain molecular fragments, information of a printed English text is created by positioning die particles on paper. Several examples of distinguishable states used to create information are shown in Table 3.1. A more detailed quantitative discussion of the concept of information is provided in Chapter 6. Also, comprehensive discussion of the topic in [3] can be recommended for additional reading.

Information of arbitrary kind and amount (such as letters, numbers, colors, or graphics specific sequences and patterns) can be represented by combination of just two distinguishable states, **0** and **1**. The maximum amount of information, which can be conveyed by a system with just two states is used as a unit of information known as a 1 bit (abbreviated from 'binary digit').

A system with two distinguishable and controllable states forms a basis for the *binary switch*, the fundamental computational element in information-processing systems (Fig. 3.1).

Three essential properties of a binary switch are *Distinguishability, Controllability* and *Communicativity*. We say that a binary switch is *Distinguishable* if and only if the binary state (0 or 1) can be

Table 3.1 Examples of distinguishable states used to create information

Information coding system	Number of Distinguishable states
English alphabet	**27: a, b, c, ...z, 'space'**
Morse code	**3:** • ('dot'), — ('dash'), **'space'**
Genetic code (DNA)	**4: A** (adenine), **C** (cytosine), **G** (guanine), **T** (thymine)
Binary code	**2: 1** and **0**

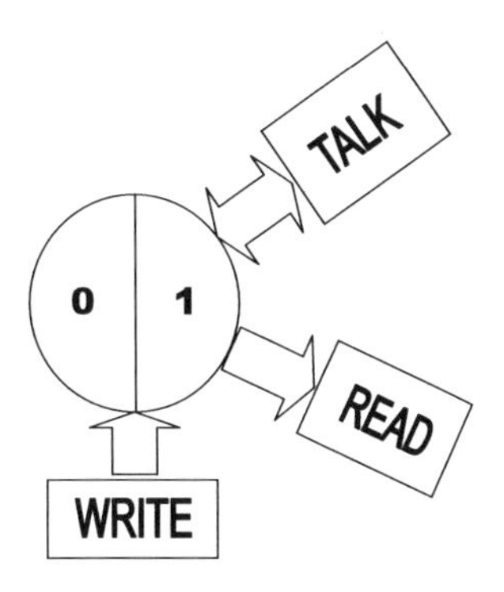

FIGURE 3.1

Constituents of an abstract binary switch

determined with an acceptable degree of certainty by a measurement (READ operation). The binary switch is *Controllable* if an external stimulus can reliably change the state of the system from 0 to 1 or from 1 to 0 (WRITE operation). The binary switch is communicative if it is capable of transferring its state to other binary switches (TALK operation).

An arbitrary binary information-processing system consists of N binary switches connected in a certain fashion to implement a specific function (e.g. logic, arithmetic etc.). Each binary switch is characterized by a dimension L and switching time t_{sw} (or switching frequency $f = 1/t_{sw}$). A related dimensional characteristic is the number of binary switches, N (or the number of binary switches per unit area, n). If area is fixed, to increase N, the characteristic dimension, L, of the binary switch must decrease:

$$N \sim \frac{1}{L^2} \tag{3.1}$$

One indicator of the ultimate performance of an information processor, realized as an interconnected system of binary switches, is the *maximum binary throughput (BIT);* that is the maximum number of binary transitions per unit time:

$$BIT = \frac{N}{t_{sw}} = N \cdot f \tag{3.2}$$

One can increase the binary throughput by increasing the number of binary switches N, and/or by decreasing the switching time, i.e., the time to transition from one state to the other, t_{sw}. Increased binary throughput has historically resulted in an increased information-processing system capability. Table 3.2 shows several examples of Intel microprocessors characterized by the number of switches (transistors), switching (clock) frequency, and their maximum binary throughput.

Another fundamental characteristic of a binary switch is the switching energy E_{sw} and the related power dissipation by a system of N binary switches is:

$$P = \frac{N}{t_{sw}} \cdot E_{sw} = BIT \cdot E_{sw} \tag{3.3}$$

In the next sections the fundamental relations for n_{bit}, t_{sw}, E_{sw} and the corresponding implications for the computing systems are investigated.

Table 3.2 Examples of Intel microprocessors with respect of the number of switches, switching frequency, maximum binary throughput and their computational capability*

Processor	# Switches (transistors)	Switching (clock) frequency	Max. binary throughput	Capability/ application
8008	3500	200 kHz	7×10^8	General calculators
8080	6000	2 MHz	1.2×10^{10}	1st PC
Pentium Extreme 965	376 000 000	3.73 GHz	1.4×10^{18}	High-performance desktop

Data from the Intel Microprocessor Quick Reference Guide (http://www.intel.com/pressroom/kits/quickreffam.htm).

BOX 3.2 INFORMATION CARRIERS

Information-processing systems represent system states in terms of physical variables. To create, change, and communicate between the states, *information carriers* are needed, and are generally *material particles* of a given kind. Examples of information carriers are *electrons, ions/atoms, photons,* etc. The corresponding physical attributes associated with information carriers could be: electrical charge (e.g. electrons or ions), spin (usually electrons), electromagnetic field (intensity and/or polarization), etc. Devices using different information carriers/state variables are often grouped in categories, such as *Electronics, Spintronics,* and *Photonics.* This chapter is mainly focused on electron-based devices (although many derivations and conclusions are universal across all information carriers). As the authors argued in [4], electronic devices (compared to spintronic and photonic devices) appear to be more suitable in systems where size and energy are primary constraints. Recently, a new class of nano-scale devices have been reported, which uses ions as information carriers [5–8]. These *nanoionic* devices may have potential for scaling beyond the limits of the electron-based devices. Interestingly, ions are believed to play a fundamental role in information processing by biological systems [9].

3.3 BASIC PHYSICS OF BINARY ELEMENTS

3.3.1 Distinguishable states

One way to create physically distinguishable states is by the *presence or absence* of material particles in a given location. Figure 3.2a shows an abstract model for a binary switch whose state is represented by different positions of a material particle. In principle, the particle can possess arbitrary mass, charge, etc. The only two requirements for the implementation of a particle-based binary switch are (i) the ability to detect the presence/absence of the particle in e.g., the location x_1, and (ii) the ability to move the particle from x_0 to x_1 and from x_1 to x_0. If it is assumed that the information-defining particle in the binary switch has zero velocity/kinetic energy, prior to a WRITE command, then it would remain where placed without constraints. However, each material particle *at equilibrium* with the environment possesses kinetic energy of $\frac{1}{2}\, k_B T$ per degree of freedom due to thermal interactions, where k_B is the Boltzmann's constant and T is temperature. The permanent supply of thermal energy to the system occurs via mechanical vibrations of atoms (phonons) and via the thermal electromagnetic field of photons (background radiation).

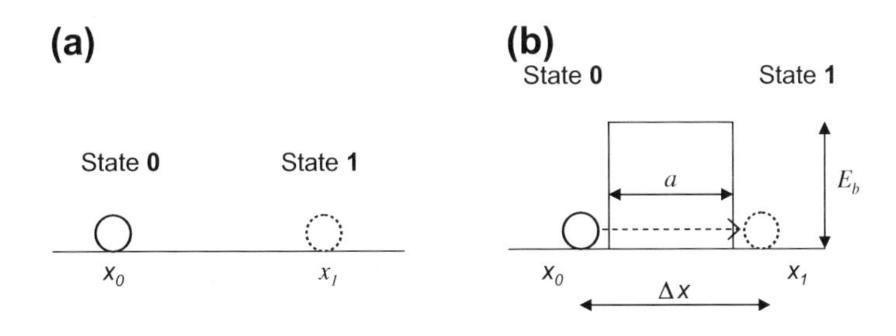

FIGURE 3.2

(a) Creating distinguishable states by different positions of a material particle and (b) energy barrier to preserve the binary states

The existence of random mechanical and electromagnetic stimuli means that the information carrier/material particle located in x_0 (Fig. 3.2a) has a non-zero velocity in a non-zero temperature environment and that it will spontaneously move from its intended location.

In order to prevent the location of the particle from changing randomly due to thermal excitation, energy barriers can be constructed that limit particle movements. The energy barrier, separating the two states in a binary switch is characterized by its height E_b and width a (Fig. 3.2b).

The barrier height, E_b, must be large enough to prevent spontaneous transitions (errors). Two types of unintended transitions can occur: 'classical' and 'quantum'. The 'classical' error occurs when the particle jumps over barrier. This can happen if the kinetic energy of the particle E is larger than E_b. The corresponding probability for over-barrier transition Π_C (referred herein as 'classic' error probability), is obtained from the Boltzmann distribution as:

$$\Pi_C = \exp\left(-\frac{E_b}{k_B T}\right). \tag{3.4}$$

Another class of errors, called 'quantum errors', occur due to quantum mechanical effects. These effects play a measurable role in a small system with energy (E), momentum (p), space (l) and time (t) scales for the processes occurring in the system are very small such that the characteristic physical parameter, the action, $S \sim E \cdot t \sim p \cdot l$, is comparable with the *quantum of action* $h = 6.63 \times 10^{-34}$ J·s, also known as *Planck's constant*. The corresponding relations are known as *Heisenberg Uncertainty Principle*:

$$\begin{aligned} \Delta x \cdot \Delta p &\sim h \\ \Delta E \cdot \Delta t &\sim h \end{aligned} \tag{3.5a}$$

The relations (3.5) are usually treated as coarse approximations with not well-defined boundaries. In certain cases, however, it is possible to define constant, α, such that the indicated relational forms in (3.5a) are more definitive:

$$\begin{aligned} \Delta x \cdot \Delta p &\geq \alpha h \\ \Delta E \cdot \Delta t &\geq \alpha h \end{aligned} \tag{3.5b}$$

In principle, there are rules, for selection of α depending on the boundary conditions [26], however elucidation of the selection criteria is beyond the scope of this text.

Typical values of α used in the literature are: 1, ½, ¼, $\frac{1}{2\pi}$, $\frac{1}{4\pi}$. For a number of model problems it is possible to determine a represented *sharpest obtainable bound* [11–13]. In this case, the relations (3.5b) can be treated as equalities, and used for quantitative estimates with reasonable accuracy. For example, the sharpest obtainable bound for the coordinate-momentum relation of a free moving electron was found to be $\alpha = \frac{1}{4\pi}$ and thus [11]:

$$\Delta x \Delta p \geq \frac{\hbar}{2} \tag{3.6}$$

where $\hbar = h/2\pi$ is the reduced Planck's constant.

Equation (3.6) will be used to estimate the limits of distinguishability due to the 'quantum errors', which occur due to quantum mechanical tunneling through a barrier of finite width a. If the barrier is too narrow, spontaneous tunneling through the barrier will destroy the binary information. The conditions for significant tunneling can be estimated using the Heisenberg uncertainty relation (3.6); as is often done in the texts on the theory of tunneling [10].

Consider again a 'two-well' bit in Fig. 3.2b. As is known from quantum mechanics, a particle can pass (tunnel) through a barrier of finite width even if the particle energy is less than the barrier height, E_b. An estimate of how thin the barrier must be to observe tunneling can be made from (3.6) as follows.

The well-known relations for kinetic energy, E, and momentum, p, of a particle of mass m and velocity v are:

$$p = mv \tag{3.7a}$$

$$E = \frac{mv^2}{2} = \frac{p^2}{2m} \tag{3.7b}$$

or

$$p = \sqrt{2mE} \tag{3.7c}$$

For a particle in the well, $E_{min} = 0$ and $E_{max} = E_b$ and corresponding momenta $p_{\min} = 0$ and $p_{\max} = \sqrt{2mE_b}$. Thus the uncertainty in momentum is:

$$\Delta p = p_{\max} - p_{\min} = \sqrt{2mE_b} - 0 = \sqrt{2mE_b} \tag{3.8}$$

From (3.6) and (3.8) we obtain:

$$\sqrt{2mE_b}\Delta x \geq \frac{\hbar}{2} \tag{3.9}$$

Equation (3.9) implies that by initially setting the particle on one side of the barrier, one can find the particle on either side, if Δx is larger than the barrier width a. Stating it differently, Δx is the uncertainty interval of particle localization, and a particle can be found anywhere within this interval. If a barrier is present within the position uncertainty interval Δx, and the barrier width a is less than Δx, the particle does not 'feel' the presence of the barrier (see Fig. 3.2b). That is, the condition for losing distinguishability is $\Delta x \geq a$, and the minimum barrier width is:

$$\Delta x \sim a_{\min} = a_H = \frac{\hbar}{2\sqrt{2mE_b}} \tag{3.10}$$

where a_H is the *Heisenberg distinguishability length* for 'classic to quantum transition'.

For $a < a_H$, tunneling probability is significant, and therefore particle localization is not possible.

To estimate the probability of tunneling, re-write (3.10), taking into account the tunneling condition $a \leq \Delta x$ to obtain:

$$\sqrt{2m}\left(a\sqrt{E_b}\right) \leq \frac{\hbar}{2} \tag{3.11}$$

1. From (3.11), we can also write the 'tunneling condition' in the form:

$$1 - \frac{2\sqrt{2m}}{\hbar}a\sqrt{E_b} \geq 0 \tag{3.12}$$

Since for small x, $\exp(-x) \sim 1 - x$, the tunneling condition then becomes:

$$\exp\left(-\frac{2\sqrt{2m}}{\hbar}\cdot a\cdot\sqrt{E_b}\right) \geq 0 \tag{3.13}$$

2. The left side of Eq. (3.13) has the properties of probability. Indeed, it represents the tunneling probability through a rectangular barrier given by the Wentzel-Kramers-Brillouin (WKB) approximation [11]:

$$\Pi_{WKB} \sim \exp\left(-\frac{2\sqrt{2m}}{\hbar}\cdot a\cdot\sqrt{E_b}\right) \tag{3.14a}$$

This equation also emphasizes the parameters controlling the tunneling process. They are the barrier height E_b and barrier width a as well as the mass m of the information-bearing particle. If separation between two wells is less than a_H, the structure of Figure 3.4b would allow significant tunneling. In fact it is instructive to examine the physical meaning of (3.10), which was marked as the condition of significant tunneling or 'classic to quantum transition'. Substituting (3.10) into (3.14a) gives an estimate for tunneling probability through a rectangular barrier of width a_H as:

$$\Pi_{WKB} \sim \exp\left(-\frac{2\sqrt{2m}}{\hbar}\cdot a_H\cdot\sqrt{E_b}\right) = \exp(-1) \approx 0.37 \tag{3.14b}$$

Thus, the Heisenberg distinguishability length a_H from (3.10) corresponds to a tunneling probability of approximately 37%.

As was discussed above, there are two mechanisms for spontaneous transitions (errors) in the binary switch: the over-barrier transition ('classic' error) and through-barrier tunneling ('quantum' error). The probabilities of the classic and quantum errors are given by (3.4) and (3.14a) respectively. Assuming the two barrier transitions are independent, the joint error probability of the two mechanisms is:

$$\Pi_{err} = \Pi_C + \Pi_Q - \Pi_C\cdot\Pi_Q \tag{3.15}$$

Or, from (3.4) and (3.14a), we obtain:

$$\Pi_{err} = \exp\left(-\frac{E_b}{k_BT}\right) + \exp\left(\frac{2\sqrt{2m}}{\hbar}\cdot a\sqrt{E_b}\right) - \exp\left(-\frac{\hbar E_b + 2ak_BT\sqrt{2mE_b}}{\hbar k_BT}\right) \tag{3.16}$$

3.3.2 Energy barrier framework for the operating limits of binary switches

Limits on barrier height

The minimum energy to effect a binary transition is determined by the energy barrier. The work required to suppress the barrier is equal to or larger than E_b. Thus, the minimum energy for a controlled binary transition is given by the minimum barrier height of a binary switch. The minimum barrier height can be found from the distinguishability condition, which requires that the probability of errors $\Pi_{err}<0.5$, in which case the switch is being operated at the threshold of distinguishability. First, consider the case when only 'classic' (i.e. thermal) errors can occur. In this case, according to (3.4), repeated here for convenience:

$$\Pi_{err} = \Pi_C = \exp\left(-\frac{E_b}{k_B T}\right) \tag{3.17}$$

These classic transitions represent the thermal (Nyquist-Johnson) noise. Solving (3.17) for $\Pi_{\text{err}} = 0.5$, obtain the Boltzmann's limit for the minimum barrier height, E_{bB} as

$$E_{bB} = k_B T \ln 2 \approx 0.7 k_B T \tag{3.18}$$

Equation (3.18) corresponds to the minimum barrier height, the point at which distinguishability of states is completely lost due to thermal over-barrier transitions. In deriving (3.18), tunneling was ignored, i.e. the barrier width is assumed to be very large, $a >> a_H$.

Next, we consider the case where only quantum (i.e. tunneling) errors can occur. In this case, according to (3.14a):

$$\Pi_{err} = \Pi_Q \sim \exp\left(-\frac{2\sqrt{2m}}{\hbar} \bullet a \bullet \sqrt{E_b}\right) \tag{3.19}$$

Solving (3.19) for $\Pi_{\text{err}} = 0.5$, obtain the Heisenberg's limit for the minimum barrier height, E_{bH} as

$$E_{bH} = \frac{\hbar^2}{8ma^2} (\ln 2)^2 \tag{3.20}$$

Equation (3.20) corresponds to a narrow barrier, $a \sim a_H$, the point at which distinguishability of states is lost due to tunneling transitions. In deriving (3.20), over-barrier thermal transitions were ignored, i.e. the temperature was assumed close to absolute zero, $T \to 0$.

Limits on size

The minimum size of a binary switch L cannot be smaller than the distinguishability length a_{H}. From (3.6) and (3.18) one can estimate the Heisenberg's length for the binary switch operation at the Boltzmann's limit of energy:

$$a_{HB} = \frac{\hbar}{2\sqrt{2mk_B T \ln 2}} \tag{3.21}$$

For electrons ($m = m_e = 9.31 \times 10^{-31}$ kg)) at $T = 300$ K obtain $a_{HB} \sim 1$ nm.

Limits on speed

The next pertinent question is the minimum switching time τ_{min}. This can be derived from the Heisenberg relation for time and energy:

$$\Delta E \Delta t \geq \frac{\hbar}{2} \tag{3.22a}$$

or

$$\tau_{\min} = \tau_H \cong \frac{\hbar}{2\Delta E} \tag{3.22b}$$

where τ_H is the *Heisenberg time,* which represents the zero-length approximation for the switching speed. For the Boltzmann's limit, $E = E_{bB}$ (3.18), obtain

$$\tau_{HB} \cong \frac{\hbar}{2k_BT \ln 2} \approx 2 \cdot 10^{-14} s \tag{3.23}$$

Note that while (3.22b) can be regarded as an imprecise upper bound for timing physical events, it also can be refined for certain model problems [12, 13]. For example, the sharpest obtainable bound for the passage time, e.g. from point **A** to point **B**, was shown to be [13]:

$$\tau_{\min} \cong \frac{h}{2\Delta E} \tag{3.24}$$

(i.e. h instead of $\hbar$ is used).

Combined effect of classic and quantum errors

Now consider the case where both thermal and tunneling transitions contribute to the errors in the operation of a binary switch. In this case, the total error probability is given by (3.16). An approximate solution of (3.16) for $\Pi_{\text{err}} = 0.5$ is

$$E_{b\ \min} = k_BT \ln 2 + \frac{\hbar^2}{8ma^2} (\ln 2)^2 \tag{3.25}$$

Equation (3.25) gives a generalized value for minimum energy per switch operation at the limits of distinguishability that takes into account both classic and quantum transport phenomena. The plot given in Figure 3.3 shows the numerical solution of Eq. (3.15) and its approximate analytical solution given by Eq. (3.25) for $\Pi_{err} = 0.5$. It is clearly seen that for $a > 5$ nm, the Boltzmann's limit, $E_{bB} = k_BT \ln 2$, is a valid representation of minimum energy barrier height, while for $a < 5$ nm, the

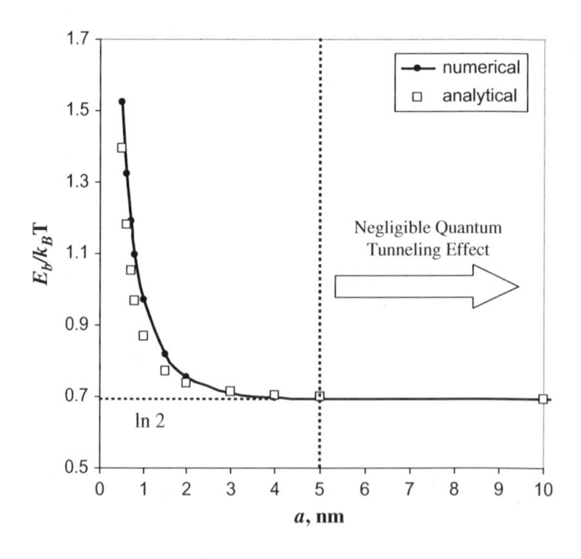

FIGURE 3.3

Minimum energy per switch operation as a function of minimum switch size minimum energy barrier height (and therefore the switching energy) must be considerably larger to retain a switching error probability of <0.5.

Main Point I:

Tunneling cannot be ignored for $a < 5$ nm, which sets a practical limit for scaling of binary switches

Quantum conductance

The Heisenberg relations can be used to evaluate the limits on electrical conductance [27].

Consider an elementary act of electrical conductance for an electron passing from reservoir **A** with energy E_A to reservoir **B** with energy E_B. The corresponding voltage (potential difference) between **A** and **B**, V_{AB} and the current, I_{AB}, flowing from **A** to **B** are:

$$|V_{AB}| = \frac{E_A - E_B}{e} = \frac{|\Delta E|}{e} \tag{3.26a}$$

$$I_{AB} = \frac{e}{\Delta t} \tag{3.26b}$$

Let Δt be the minimum passage time (3.24):

$$\Delta t = \frac{h}{2\Delta E} = \frac{h}{2eV} \tag{3.27}$$

Putting (3.27) into (3.26b), and taking into account Ohm's law, i.e. $I = V/R$, we obtain:

$$I_{AB} = \frac{2e^2}{h} \bullet V = \frac{V}{R_0} \tag{3.28}$$

where

$$R_0 = \frac{h}{2e^2} = 12.9k\Omega \tag{3.29a}$$

is quantum resistance. A related parameter is quantum conductance:

$$G_0 = \frac{1}{R_0} = \frac{2e^2}{h} \tag{3.29b}$$

The quantum resistance/conductance sets the limit on electrical conductance in a one-electron channel *in the absence of barriers.*

$$I_0 = \frac{V}{R_0} = \frac{V}{12.9k\Omega} \tag{3.30}$$

An example of practical realization of the one-electron conductance channel is single-atom contact formed by 1-valence atoms such as Cu, Ag, or Au. Indeed, experiments with the single-atom contacts demonstrated the minimum contact resistance of 12.9 kΩ for gold [28]. Multi-valence atoms can have several electrons available for conductance, thus forming several parallel conductive channels, thereby increasing conductance. For example, experiments with single-atom contact of alumimum atoms (three valence electrons) showed three times lower contact resistance than for the gold atom (one valence electron) [28]. Another way to increase the number of parallel channels and thus decrease the resistance is to increase the number of conductive atoms, i.e. increase the cross-sectional area of the conductor.

If a barrier is present in the electron transport system, the conductance will be decreased due to the barrier transmission probability $\Pi < 1$. The electrical conductance in the presence of barrier is obtained by multiplying the barrier-less quantum conductance (3.29b) by the barrier transmission probability Π:

$$G = \frac{1}{R} = G_0 \cdot \Pi \tag{3.31}$$

Equation (3.31) is a form of the *Landauer formula* [29] for a one-electron conductive channel.

BOX 3.3 ENERGY BARRIERS

The physical implementation of an energy barrier depends on the choice of the information carrier (see Box 3.2). In all cases, the energy barrier creates a local change of the potential energy of a particle. For example, for electrons, the energy barrier formed due to a change of electric potential energy of a particle from a value $e\varphi_1$ at the coordinate x_1 to a larger value $e\varphi_2$ at the coordinate x_2 as shown in Figure B3.3a (e is the electron charge, and φ is the electric potential). The difference $e\Delta\varphi = e(\varphi_2 - \varphi_1)$ is the barrier height E_b. In a system with an energy barrier, the force exerted on a particle by the barrier is of the form $F = e\frac{\partial\varphi}{\partial x}$. A simple illustration of a one-dimensional barrier in linear spatial coordinates, x, is shown in Figure B3.3a.

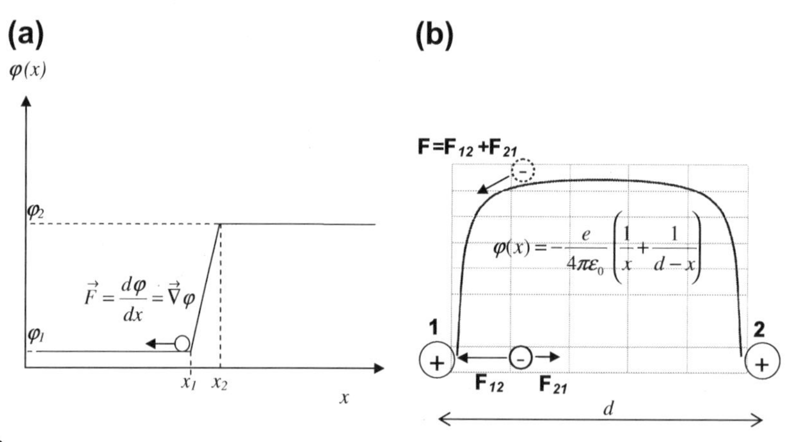

FIGURE B3.3

An illustration of the energy barrier in a material system: (a) abstraction; (b) a physical implementation example using fixed charges

As an example, consider a barrier for an electron (negatively charged) formed by two fixed positive charges separated by a distance d, as shown in Figure B3.3b. The barrier is a result of electrostatic (Coulombic) forces acting on the electron.

BOX 3.4 ENERGY BARRIERS IN MATERIAL SYSTEMS

Energy barriers (for electrons) in material systems are commonly formed at the interfaces between two materials with different concentration of electrons. Examples are shown in Figure B3.4:

(a) The barrier at the metal–vacuum interface (also called the workfunction) has a typical height of 4–5 eV for stable metals. The transition region from low to high barrier energies is about 1 nm.

(b) The barrier at the contact between two different metals is characterized by the contact potential difference (CPD). A typical barrier height is ~100 μeV, and the size of barrier region is ~0.1 nm (for more discussion see Box 4.5 in Chapter 4).

(c) A barrier between a metal and a semiconductor, also called *Schottky barrier* has a typical barrier height of ~0.5–1 eV (for Si). The size of the barrier region (depletion length) W depends on the concentration of carriers in semiconductors; typically $W \sim$ 10–1000 nm

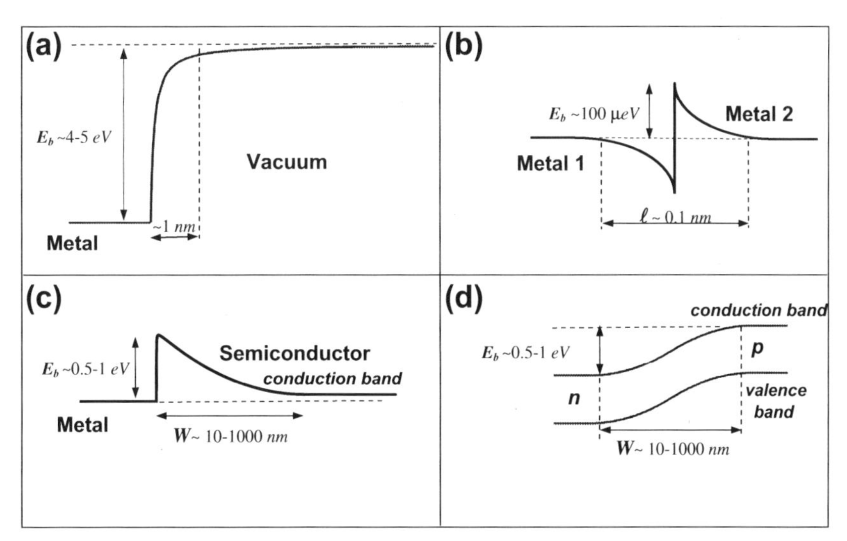

FIGURE B3.4

Energy barriers in material systems: (a) metal–vacuum; (b) metal–metal, (c) Schottky barrier; (d) pn-junction

(d) A barrier is formed between two regions of a semiconductor with different types of conductivity, also called *pn-junction* (more details are provided in Chapter 5). The typical barrier height is ~ 0.5–1 eV (for Si); the size of the barrier region (depletion length) W depends on the concentration of carriers in semiconductors; typically $W \sim$ 10–1000 nm.

3.3.3 A summary of device scaling limits

Based on the arguments in the previous sections, an arbitrary binary device can be represented by a generic barrier model of Figure 3.4. The energy barrier is needed to preserve a binary state in the presence of classic (thermal) and quantum (tunneling) errors (noise). The barrier properties, namely barrier height, E_b, and barrier width, a, determine the lower bound on the operational energy and size of binary device, as it is summarized in Table 3.3.

BOX 3.5 SEMICONDUCTOR FIELD EFFECT TRANSISTOR

The energy barrier for electrons can be formed by built-in charges placed in a material system, an approach utilized in semiconductor devices. When certain impurity atoms are introduced into a matrix of semiconductor material (a process known as doping), these atoms spontaneously ionize, thus forming built-in positive or negative charges in the semiconductor matrix. For example, the atoms of phosphorus (**P**), when introduced into silicon (**Si**), are charged positively (**P^+**), while the atoms of boron (**B**) in Si are charged negatively (**B^-**). The concentration of impurities in the semiconductor matrix is very small (typically 0.002–200 ppm). A combination of silicon regions doped with **P** and **B** shown in Figure B3.4a results in a barrier structure shown in Figure B3.4b. In order to control the barrier height in the semiconductor structure, a metal electrode (gate) is coupled to the barrier region, separated by a thin layer of insulator. When a voltage V_g is applied to the gate, an electric field is created between the gate and the barrier region. This electric field changes the barrier height, thus allowing electrons to pass. The structure of Figure B3.3 is used in devices called field effect transistors (**FET**) and is represented symbolically as shown in Figure B3.4C.

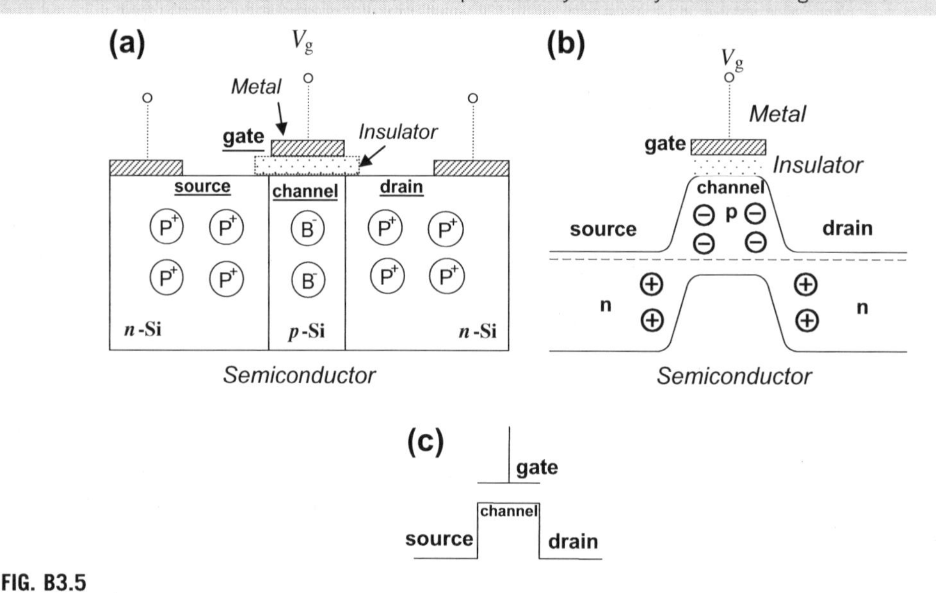

FIG. B3.5

Semiconductor field effect transistor: (a) materials system; (b) barrier representation; (c) schematic symbol

The generic materials component of FET are: a *metal* (gate), a gate *insulator* (separates the gate from the semiconductor region), and a *semiconductor* (where the barrier is formed). The name for the device is MIS FET. In most practical cases, the insulator is made of oxide (SiO_2, and more recently HfO_2), and the most common name for this class of devices is **MOSFET**. Silicon MOSFET binary switches form the platform for modern digital electronics.

3.3.4 Charge-based binary logic switch

In this section, an analysis is offered for the electron-charge-based binary switch. For electrons, the basic equation for potential energy is the Poisson equation:

$$\frac{d^2\varphi}{dx^2} = -\frac{\rho}{\varepsilon_0}, \tag{3.32}$$

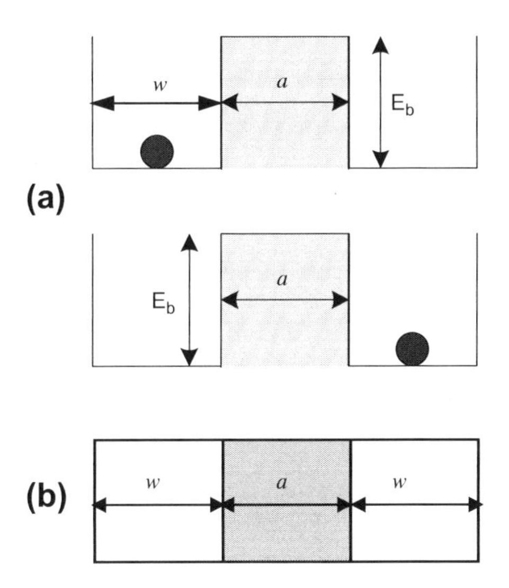

FIGURE 3.4

A barrier model for a binary switch: (a) schematic energy diagram; (b) generic floorplan

Table 3.3 Energy–space–time device boundaries

Boltzmann limit	Heisenberg limit	
E_{min}	L_{min}	t_{min}
$\Pi_{err} = \exp\left(-\frac{E_b}{k_B T}\right)$	$\Delta x \Delta p \geq \frac{\hbar}{2}$	$\Delta E \Delta t \geq \frac{\hbar}{2}$
$E_b^{min} = k_B T \ln 2$	$a_H = \frac{\hbar}{2\sqrt{2mE_b}}$ [1]	$\tau_H = \frac{\hbar}{2\Delta E}$ [2]
Boltzmann-Heisenberg limit[3]		
$E_B \sim 10^{-21}$ J	a_{HB}~1 nm	$t_{HB} \sim 0.02$ ps

[1] a_H *is the Heisenberg distinguishability length for 'classic to quantum transition',* m = m_e*, the free electron mass.*
[2] τ_H *is the Heisenberg time, which represents the zero-length approximation for the switching speed.*
[3] a_{HB} *and* τ_{HB} *are, respectively, the Heisenberg length and time calculated for the Boltzmann's limit,* $E_b = k_B T \ln 2$.

where ρ is the charge density, $\varepsilon_0 = 8.85 \times 10^{-12}$ F/m is the permittivity of free space, and φ is the electric potential. According to (3.32), the presence of an energy barrier is associated with changes in charge density in the barrier region. As discussed in Box 3.5, the barrier-forming charges can be introduced in a semiconductor material by doping (this is illustrated in Fig. B3.5, for a doped silicon structure where the barrier is formed by ionized impurity atoms such as $\mathbf{P}^+$ and $\mathbf{B}^-$). The barrier height E_{b0} depends on the concentration of the ionized impurity atoms [15]:

$$E_{b0} \approx k_B T \ln \frac{N_a^- N_d^+}{n_i} \tag{3.33}$$

where, N_a^- is the concentration of negatively charged impurities (acceptors, e.g. $\mathbf{B}^-$), N_d^+ is the concentration of positively charged impurities (donors, e.g. $\mathbf{P}^+$), and n_i is the 'intrinsic carrier concentration' in an undoped semiconductor (for Si, $n_i = 1.45 \times 10^{10}$ cm^{-3}).

The energy diagram of Figure B3.5b is typical for many semiconductor devices, for example, field effect transistors (FET). The barrier region corresponds to the FET channel that lies beneath the gate, while the wells correspond to the source and drain. To enable electron movement between the source and drain, the barrier height must be decreased (ideally suppressed to zero). To do this, the amount of charge in the barrier region needs to be changed, according to (3.32). A well-known relation connects the electrical potential difference $\Delta\varphi = V$ and charge, Δq, through capacitance:

$$C = \frac{\Delta q}{\Delta \varphi} \tag{3.34}$$

In field effect devices, in order to change charge distribution in the barrier region, and hence lower the barrier, a voltage is applied to an external electrode (gate), which forms a capacitor with the barrier region (in bipolar devices external charge is injected into the barrier region to control the barrier height). When voltage V_g is applied to the barrier region, the barrier will change from its initial height E_{b0} (determined by impurity concentration – Eq. 3.33):

$$E_b = E_{b0} - eV_g \tag{3.35}$$

The voltage needed to suppress the barrier from E_{b0} to zero (the threshold voltage V_t) is:

$$V_t = \frac{E_{b0}}{e} \tag{3.36}$$

Thus operation of all charge transport devices involves charging and discharging capacitances to change barrier height, thereby controlling charge transport in the device. When a capacitor C is charged from a constant voltage power supply V_g, the energy E_{dis} is dissipated, i.e. converted into heat (see Box 3.6 for derivation and [16] for further discussions):

$$E_{dis} = \frac{CV_g^2}{2} \tag{3.37}$$

The minimum energy needed to suppress the barrier (by charging the gate capacitor) is equal to the barrier height E_b. Restoration of the barrier (by discharging gate capacitance) also requires a minimum energy expenditure of E_b. Thus the minimum energy required for a full switching cycle is at least $2E_b$.

Note that in solid state implementation of binary switch, the number of electrons in both wells is usually large. The electrons strike the barrier from both sides, and the binary transitions are determined by the net electron flow, as is shown in Figure 3.5. Let N_{e0} be the number of electrons that strike the barrier per unit time. The expected number of electrons N_{eA} that transition over the barrier from well **A** per unit time is:

$$N_{eA} = N_{e0} \exp\left(-\frac{E_b}{k_B T}\right) \tag{3.38}$$

The corresponding current I_{AB} is:

$$I_{AB} = e \bullet N_{eA} = eN_{e0} \exp\left(-\frac{E_b}{k_B T}\right) \tag{3.39}$$

BOX 3.6 ENERGY DISSIPATION BY CHARGING A CAPACITOR

Energy dissipation by charging a capacitor is a central concept of microelectronics. In fact, operation of all electronic devices involves charging/discharging corresponding capacitors. When a capacitor is charged from a constant voltage power supply, the energy is dissipated, i.e. converted into heat. Consider a typical model circuit consisting of a capacitor C in series with a resistor R (Fig. B3.6). Suppose a constant voltage of magnitude V is applied to the circuit at $t = 0$ and electrical charge flows to the capacitor. The charging of the capacitor is characterized by a time-dependent voltage drop both on the resistor and the capacitor:

$$V_R(t) = V\exp\left(-\frac{t}{RC}\right) \tag{B.3.6.1a}$$

$$V_C(t) = V\left(1 - \exp\left(-\frac{t}{RC}\right)\right) \tag{B3.6.1b}$$

The energy dissipated in the resistor R during charging is:

$$E_R = \int_0^\infty \frac{V_R^2(t)}{R}dt = \frac{V}{R}\int_0^\infty \left[\exp\left(-\frac{t}{RC}\right)\right]^2 dt = \frac{CV^2}{2} \tag{B3.6.2}$$

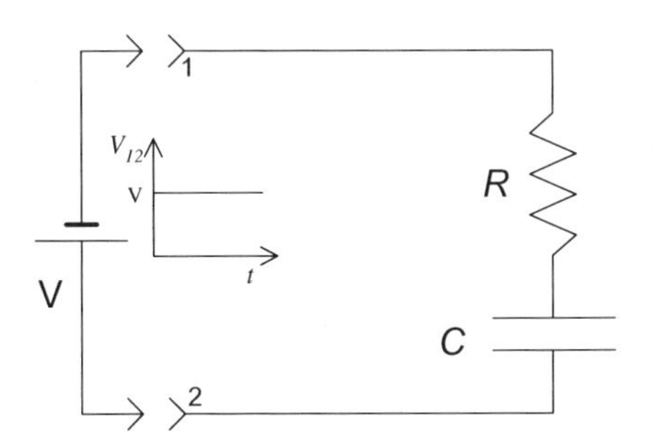

FIGURE B3.6

Generic RC circuit

Note that the energy dissipated in the resistor is independent of the resistance value R. As a result of the charging process, the capacitor stores the energy $E_C = ½CV^2$, and thus the total energy required for a constant charging voltage is CV^2.

Electrons in well **B** also can strike the barrier and therefore contribute to the over-barrier transitions with current I_{BA}. Thus the net over-barrier current is:

$$I = I_{AB} - I_{BA} \tag{3.40}$$

The energy diagram of Figure 3.5a is symmetric, hence $I_{AB} = I_{BA}$, and $I = 0$. Therefore, no binary transitions occur in the case of the symmetric barrier. In order to enable rapid and reliable transition of an electron from well **A** to well **B**, an energy asymmetry between two wells must be created. This is achieved by energy difference eV_{AB} between the wells **A** and **B** (Fig. 3.5b).

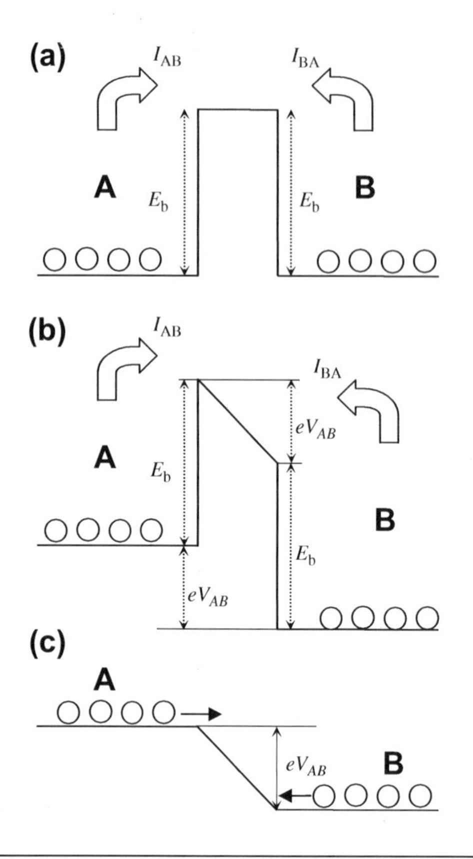

FIGURE 3.5

Fundamental operation of multi-electron binary switch: (a) no energy difference between the wells **A** and **B** resulting in a symmetric energy diagram; (b) Energy asymmetry is created due to energy difference eV_{AB} between the wells **A** and **B**; (c) state CHANGE operation: the barrier height E_b is suppressed by applying gate potential V_g, while energy difference eV_{AB} between the wells **A** and **B** is maintained

For such an asymmetric diagram, the barrier height for electrons in the well **A** is E_b, and for electrons in the well **B** is $(E_b + eV_{AB})$. Correspondingly, from (3.39) and (3.40) the net current is

$$I = eN_{e0}\exp\left(-\frac{E_b}{k_BT}\right) - eN_{e0}\exp\left(-\frac{E_b - eV_{AB}}{k_BT}\right) = eN_{e0}\exp\left(-\frac{E_b}{k_BT}\right)\left[1-\exp\left(-\frac{eV_{AB}}{k_BT}\right)\right] \tag{3.41a}$$

Substituting (3.35) for E_b we obtain:

$$I = eN_{e0}\exp\left(-\frac{E_{b_0} - eV_g}{k_BT}\right)\left[1-\exp\left(-\frac{eV_{AB}}{k_BT}\right)\right] \tag{3.41b}$$

Expressing E_{b0} as eV_t from (3.36) and using the conventional notations $I = I_{ds}$ (source-drain current) and $V_{AB} = V_{ds}$ (source-drain voltage) we obtain the equation for subthreshold I–V characteristics of FET [17]:

$$I_{ds} = I_0 \exp\left(\frac{e(V_g - V_t)}{k_B T}\right)\left[1 - \exp\left(-\frac{eV_{ds}}{k_B T}\right)\right] \quad (3.41c)$$

The *minimum* energy difference between the wells, eV_{ABmin}, can be estimated based on the distinguishability arguments for CHANGE operation, when E_b is suppressed by applying the gate voltage, e.g. $E_b = 0$ (Fig. 3.5c). For a successful change operation, the probability that each electron flowing from well **A** to well **B** is *not* counterbalanced by another electron moving from well **B** to well **A** should be less than 0.5 in the limiting case. The energy difference eV_{AB} forms a barrier for electrons in well **B**, but not for electrons in well **A**, therefore, from (3.4) we obtain:

$$eV_{AB_{min}} = E_{b_{min}} = k_B T \ln 2 \quad (3.42)$$

If N_e is the number of electrons involved in the switching transition between two wells, the total minimum switching energy is

$$E_{SW_{min}} = 2E_b + N_e eV_{AB} = (N_e + 2) k_B T \ln 2 \quad (3.43a)$$

If $N_e = 1$,

$$E_{SW_{min}} = 3k_B T \ln 2 \approx 10^{-20} J \quad (3.43b)$$

3.3.4 Charge-based memory element

A digital memory element, in its most fundamental form, consists of:

1. Two states 0 and 1, which are equally attainable and distinguishable.
2. A means to control (change) the state.
3. A means to read the state.

In electron-based memory, the two distinguishable states are created by the presence (e.g. STATE 0) or absence (e.g. STATE 1) of electrons in a specific location, called the *storage node*.

A material structure to implement an electron-based memory element must satisfy two requirements: (1) *Charge retention*: very small flow of charge in store mode (practical requirement for the retention time of non-volatile memory is $t_r \sim 10$ years) and (2) *Charge injection*: Sufficiently large flow of charge during write mode. In the store mode the charge must be confined within a storage node, whose dimensions define the size of the memory element. In order to prevent charge losses, the storage node is defined by energy barriers of sufficient height E_b to retain charge (as shown in Fig. 3.6). Barrier parameters E_b and a (assumed symmetrical) are the critical considerations for memory operation, as will be discussed below. The two-barrier diagram of Figure 3.6 serves as a generic abstraction for all types of electron charge-based memories, for example: DRAM, floating gate (Flash/SONOS), etc.

Assume the memory element stores one electron. The corresponding characteristic one-electron escape time $\boldsymbol{t_{s1}}$ is:

$$t_{s1} = \frac{e}{I_s} \quad (3.44)$$

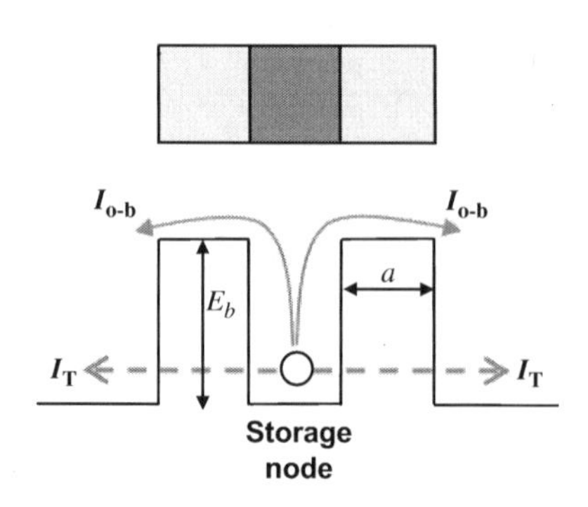

FIGURE 3.6

A generic abstraction for a charge-based memory element

The two mechanisms of the charge loss are over-barrier leakage and through-barrier tunnel leakage. In both cases the leakage current from the storage node can be calculated from the Landauer formula (3.31):

$$I_s = G_0 \cdot V \cdot \Pi \tag{3.45}$$

In (3.45) G_0 is the quantum conductance, $G_0 = \frac{2e^2}{h}$, Π is the barrier transmission probability, and $V = \frac{k_B T}{2e}$ is the thermal voltage.

The probability of thermal overbarrier transitions is the Boltzmann probability (3.4). From (3.45) and (3.4) the one-electron over-barrier current $I_{o\text{-}b}$ is:

$$I_{o-b} = 2\frac{e}{h} \cdot k_B T \cdot \exp\left(-\frac{E_b}{k_B T}\right) \tag{3.46}$$

The factor of 2 in (3.46) appears because escape is possible over either of two barriers that confine an electron as shown in Figure. 3.6.

The electron escape time (the retention time) due to over-barrier transport is:

$$t_{o-b} = \frac{h}{2k_B T} \exp\left(\frac{E_b}{k_B T}\right) \tag{3.47}$$

If over-barrier leakage is the only mechanism of charge loss (when the barrier width a is sufficient to suppress tunneling), the escape time is equal to the one-electron retention time, $t_{o\text{-}b} = t_r$.

For a specified t_r, the required minimum barrier height is:

$$E_{b\ \min} = k_B T \ln\left(\frac{2k_B T}{h} t_r\right) \tag{3.48}$$

In the case of the 'minimum non-volatile memory' requirement, i.e. $t_r > 10$ years (3.48), gives $E_{bmin} \geq 1.29$ eV at $T = 300$ K.

A second source of charge loss is electron tunneling. The tunneling probability is given by (3.14a) and the corresponding tunneling current I_T is:

$$I_T = 2\frac{e}{h} \cdot k_B T \cdot \exp\left(-\frac{2\sqrt{2m}}{\hbar} \cdot a \cdot \sqrt{E_b}\right) \tag{3.49}$$

The electron escape time due to tunneling is:

$$t_T = \frac{h}{2k_BT}\exp\left(\frac{2\sqrt{2m}}{\hbar}\bullet a\bullet\sqrt{E_b}\right) \tag{3.50}$$

The total retention time due to both mechanisms can be estimated as:

$$t_r = \frac{e}{I_{o-b} + I_T} \tag{3.51}$$

Suppose that the barrier height is large enough to suppress over-barrier escape, i.e. $E_b > E_{bmin}$, where E_{bmin} is given by (3.48). In this case, the retention time will be determined by the tunneling time, t_T: $t_r \approx t_T$. The minimum barrier width for a specified retention time, can be estimated from (3.50), e.g. for $t_r = 10$ years:

$$a_{\min} = \frac{\hbar}{2\sqrt{2mE_b}}\ln\frac{k_BT}{h}t_r \tag{3.52}$$

As a numerical estimate for $t_r > 10$ years, $E_{bmin} \geq 1.29$ eV, $m = m_e$ and $T = 300$ K (3.52) gives $a_{\min} \sim 5$ nm.

Note that in Eqs. (3.46–3.52) a simple sum of escape probabilities was used. Strictly speaking, for a one-electron case, Eq. (3.15) should be used. However, for sufficiently large barriers (e.g. barriers required for 10 years retention) the simple sum of probabilities is a good approximation.

Main Point II:

To satisfy the practical non-volatility requirement (i.e. retention time $\geq$ 10 years) the characteristic dimension of an electron-based memory element $a > 5$ nm, which is an estimate for the practical limit for scaling of electron charge-based memory elements

3.4 SYSTEM-LEVEL ANALYSIS

3.4.1 Tiling considerations: Device density and speed

The binary switch, represented by a one-barrier-and–two-wells energy diagram of Figure 3.4a also suggests a generic topology for the ultimately scaled device, shown in Figure 3.4b. While the smallest barrier size is limited by tunneling, it can be shown (see Appendix 1) that the smallest well size, w, is limited by the quantum confinement, and in the limiting case, $w_{min} \sim a_{HB} \sim a$. Thus, the two-dimensional floorplan of a smallest possible binary switch is a $3a \times a$ rectangle consisting of 3 *square tiles* of the same size a. The most compact layout for an array of devices must allow at least one tile between each device for insulation as shown in Figure 3.7. The tiling representation of binary switches allows one to calculate the maximum theoretical packing density of binary switches on a 2D plane [18,19]:

$$n_{\max} = \frac{1}{8a^2}. \tag{3.53}$$

From the values of a given in Table 3.3, the maximum theoretical device density $n_{\max} \sim 10^{13}$ cm^{-2} is obtained.

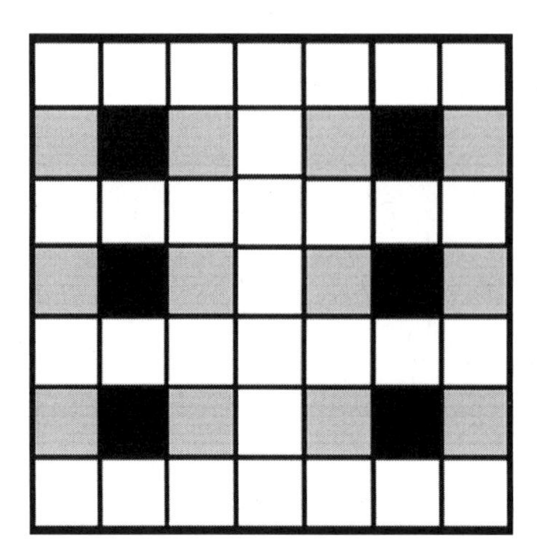

FIGURE 3.7

Most compact device layout (Joyner tiling [18,19])

As was shown at the end of Section 3.3.2, tunneling cannot be ignored at barrier width $a \leq 5$ nm (Fig. 3.3), which is consistent with the Semiconductor Roadmap projections for ultimate scaling [14]. Also, the typical packing density of transistors in practical microprocessors is less than (3.53) and, based on the ITRS data [14], it can be approximated as:

$$n_{MPU} \sim \frac{1}{(20a)^2} \tag{3.54}$$

For $a = 5$ nm, $n_{MPU} \sim 10^{10}$ cm^{-2}.

The next pertinent question is the minimum switching time. Note that in physical systems, the Heisenberg speed limit is approachable only in 'dimensionless' systems, i.e. where the material particle is moving a distance not exceeding the Heisenberg length. It is straightforward to show (see Appendix 2) that if the travel distance L is larger than a_{H}, the minimum travel time is increased as

$$\tau \sim \frac{L}{a_H} \cdot \tau_H \tag{3.55}$$

In most of this chapter, the focus is on 2D layout of electronic circuits. Indeed, 2D (planar) technology is representative of current practice. However, 3D systems are emerging, and will be necessary for the nanomorphic cell. In the following sections, a snapshot of the 3D layout for logic and memory is given.

3D Tiling: Logic

The tiling approach outlined above can be extended to the 3D case (Fig. 3.8). The three-dimensional layout of a smallest possible binary switch is a $3a \times a \times a$ cuboid consisting of 3 *cube tiles* of the same size a (Fig. 3.8a). The most compact layout for a 3D array of devices must allow at least one insulating ('white') tile between each device as shown in a top view by Figure 3.8b. The corresponding minimal insulated switch has dimensions of $4a \times 2a \times 2a$. Next, in a functional circuit interconnects are needed between the switches, and assuming the minimum interconnect metal tile of size a, the space occupied

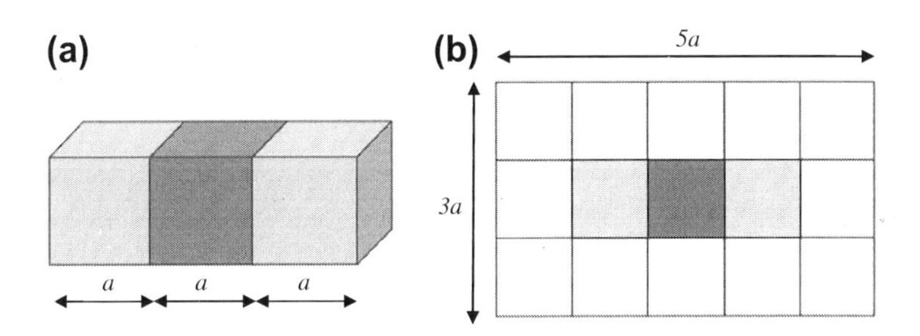

FIGURE 3.8

A generic 3D layout of the binary switch (a) and its top view (b)

by one switch in a functional circuit is $5a \times 3a \times 3a$. Thus the maximum density that could be expected for 3D organization of a functional circuit of binary switches for $a \sim 5$ nm is

$$n_{3D} \sim \frac{1}{5a \times 3a \times 3a} = \frac{1}{45a^3} \sim 10^{17} \frac{bit}{cm^3} \tag{3.56}$$

3D Tiling: Memory

The memory device is generally more complex than the two-barrier storage node structure shown in Figure 3.6. In addition to the storage node, there is also a *sensor,* which reads the state, i.e. the presence or absence of an electric charge in the storage node. The sensor is typically an electrical charge/voltage-sensitive device such as transistor, and in the limit it can be represented by a binary switch of Figures 3.4 and 3.8. Also, in a typical memory system, the memory devices (cells) are connected to form an array, and individual cells in the array are selected for read or write operations. In order to do this, memory devices in the array are located at the point of intersection of selection line ***x*** (the *bitline* – BL) and selection line ***y*** (the *wordline* – WL). These lines form contacts to the memory elements, which need to be included in the total volume estimate of the memory cell.

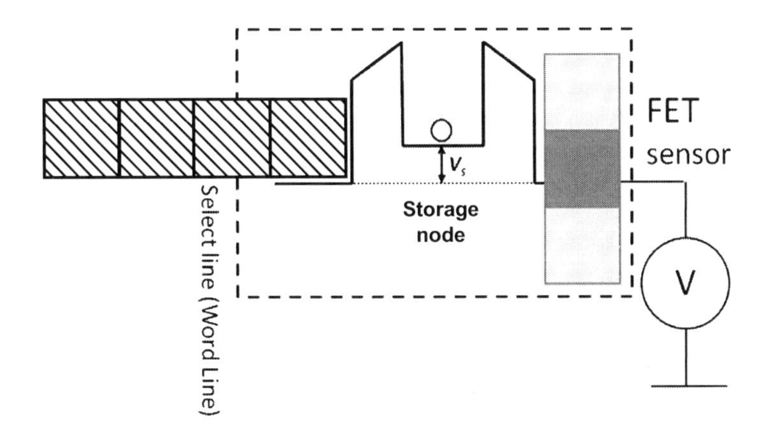

FIGURE 3.9

A generic charge-based memory element including, storage node, sensor and selector components

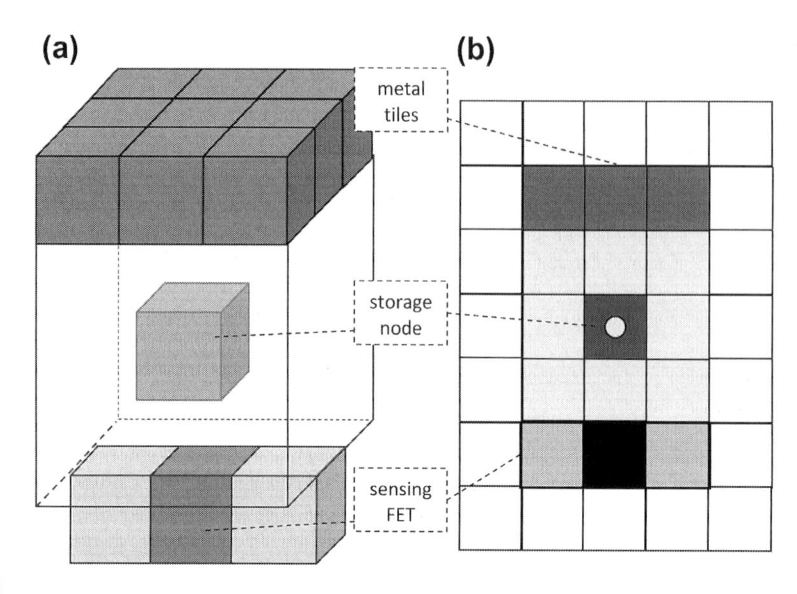

FIGURE 3.10

(a) 3D tiling layout of generic charge-based memory element, (b) the corresponding side view

Space occupied by one memory element in a 3D array can be estimated using 3D tiling similarly as was done previously for the binary switch. From Figure 3.10, the cell volume (including the necessary insulating and metal tiles) is $7a \times 5a \times 5a$ and thus the maximum density that could be expected for 3D memory array for $a \sim 5$ nm is:

$$n_{3D} \sim \frac{1}{7a \times 5a \times 5a} = \frac{1}{175a^3} \sim 10^{16} \frac{bit}{cm^3} \tag{3.56}$$

3.4.2 Adjustment for system reliability

The energy to process one bit in practical devices is higher than $k_B T \ln 2$ due to strong reliability requirements (very low error probability Π, and therefore larger E_b) and a large number of electrons, N_{el}, involved in each switching event. The three main factors that determine practical values for E_b and N_{el}, are system reliability, fan-out, and long communication requirements, as discussed below.

The requirement that all N devices in the logic system operate correctly raises E_b higher than E_{bmin} (3.18). The probability Π_{syst} of a correct operation of all N switches in a circuit is:

$$\Pi_{syst} = (1 - \Pi_{err})^N \tag{3.57}$$

where it is assumed that switch errors are independent.

From (3.57) for Π_{err} given by (3.16), one can calculate device E_b required for reliable system operation:

$$E_b = \ln\left(1 - \sqrt[N]{\Pi_{syst}}\right)^{-1} \tag{3.58}$$

Table 3.4 E_b for different system reliabilities and for different number of devices (in the units of k_BT)

N	$E_b(\Pi = 0.5)$	$E_b(\Pi = 0.99)$
1	0.69	4.61
10	2.70	6.90
100	4.98	9.21
1000	7.27	11.51
10^4	9.58	13.81
10^5	11.88	16.11
10^6	14.18	18.42
10^7	16.48	20.72
10^8	18.79	23.02
10^9	21.09	25.32
10^{10}	23.39	27.63

The values of E_b (in the units of k_BT) for different system reliabilities (correctness probabilities) and for different numbers of devices are shown in Table 3.4.

It is interesting to observe that the incremental energy cost in going from 50% to 99% reliability is not large. Therefore, the failure threshold analysis offered herein provides reasonable estimate for lower limits of energy–space–time characteristics of an information-processing system.

3.4.3 Models for connected binary switches

Devices must communicate with each other to support computation, and there is an energy cost associated with communication. In electron-based devices (e.g. MOSFET) this implies that when the electron passes from state '0' to state '1' in one binary device (sending), it needs to activate several downstream devices (receiving).

Juxtaposed switches

In a typical digital circuit, the barriers (gates) of these receiving devices are electrically coupled to a well of the first (upstream) device. Consider first the case of communication between two near-neighbor 'stacked' or juxtaposed switches as shown in Figure 3.11. When the information carrier charge from the first (upstream) device is moved from state '0' to state '1', it changes the barrier height of the downstream switch, and therefore enables the switching from '0' to '1' in the second device also.

The 'stacked' configuration of Figure 3.11 requires also 'stacked' 3D physical layout. If all binary switches are located in one plane (planar layout is employed in modern integrated circuits), an additional conducting line is needed to connect the well of the upstream device to the barrier (gate) of the downstream device. In practice, this extension is achieved by interconnect systems (Fig. 3.12). Note that the interconnect system can be represented as a combination of the square tiles in the plane. It is straightforward to show from both topological and physical considerations that in the limiting case, the size of the interconnect tile is equal to the device tile a. In the following, the plane is tiled with squares of size a.

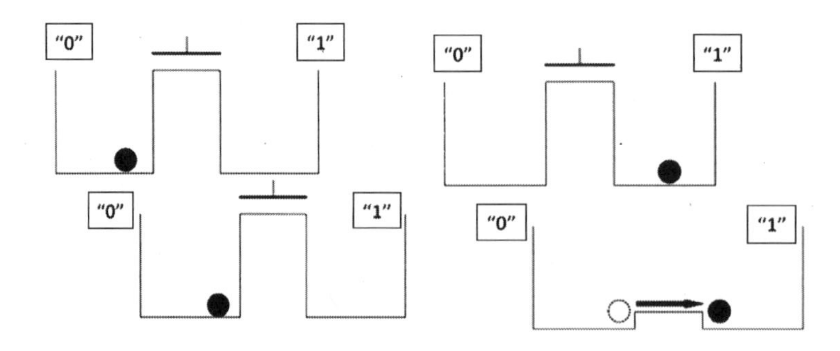

FIGURE 3.11

Communication between two 'stacked' binary switches: the charge in the "1" well of the upstream device controls the barrier height of the downstream device

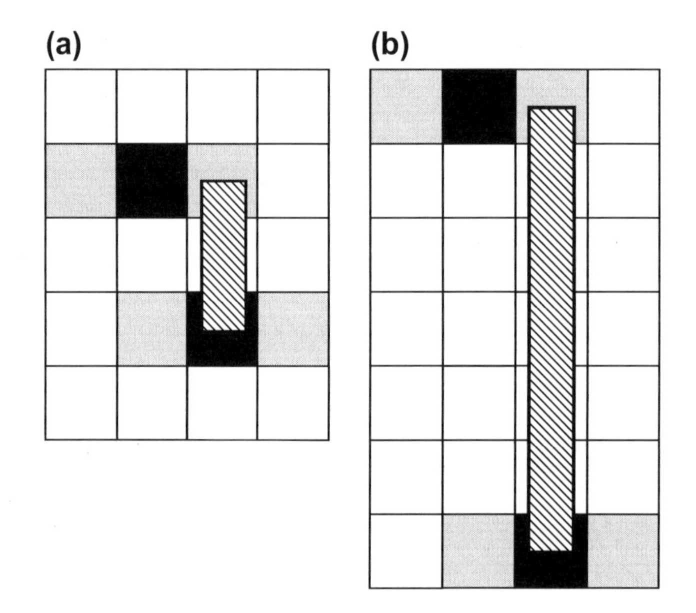

FIGURE 3.12

Tiled planar layout of interconnected binary switches: (a) minimum-length connection, and (b) a long connection

Connecting binary switches via wires: Extended well model

The junction of a conductor (e.g. metal wire) to the gate or well of a binary switch is ideally barrierless (i.e. ideal ohmic). Thus, a device well connected to a wire can be represented by *the extended well model*, where the well size is equal (approximately) to the wire length L.

The question of how reliably the charge in the extended well of the sending device **A** in Figure 3.13 controls the receiving device **B** is next addressed. Assume that one electron is needed to control the barrier of the receiving device, thus one electron passes from the well '0' to the extended well '1' in

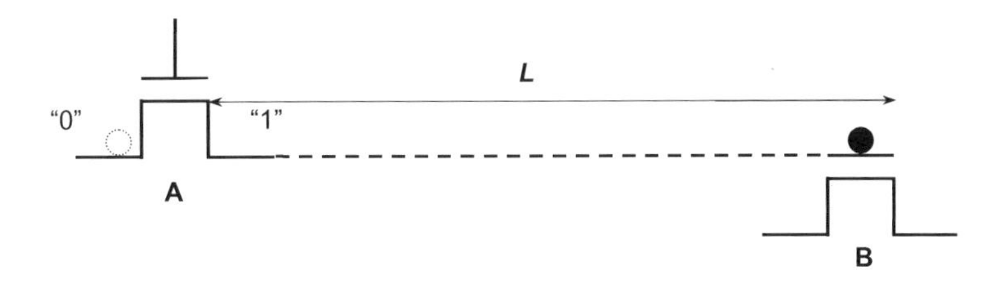

FIGURE 3.13

Extended well model for two communicated binary switches

0–1 switching of the sending device **A** ($N_e = 1$). The electron is not localized and it can freely move along the line of length L and the probability to find this electron only at gate **B** is given by:

$$\Pi_B = \frac{a}{L} \tag{3.59a}$$

and the communication error probability is

$$\Pi_{err} = 1 - \Pi_B = 1 - \frac{a}{L} \tag{3.59b}$$

For example, if $L = 4a$, the probability that the 'messenger' electron is located within the gate **B** $\Pi_B = 0.25$. This effect is known in technical literature as shot noise, which is signal fluctuation due to the discrete nature of electrical charge. In order to increase the probability of successful communication, the number of electrons must be increased. If N_e electrons are added to fill the 'pool', and these electrons move independently of each other (which is the case for electrons in metals), the communication error probability is

$$\Pi_{err}(N_e) = \left(1 - \frac{a}{L}\right)^{N_e} \tag{3.60}$$

Equation (3.60) gives the probability that *no* N_e electrons are located within the gate **B**.

Note that the number of tiles, k, needed to form an interconnect system of length L is:

$$k = \frac{L}{a} \tag{3.61}$$

To increase the probability of successful control, the number of electrons, N_e, in the interconnect line must be increased and this results in the probability of placing an electron on downstream gate **B** as:

$$\Pi_B = 1 - \left(1 - \frac{a}{L}\right)^{N_e} = 1 - \left(1 - \frac{1}{k}\right)^{N_e} \tag{3.62}$$

The solution of (3.62) for N_e is:

$$N_e = \log_{1-\frac{1}{k}}(1 - \Pi) = \frac{\ln(1 - \Pi)}{\ln\left(1 - \frac{1}{k}\right)} \tag{3.63}$$

The number of electrons N_e needed for communication between two binary switches connected by a wire of length L is given in Table 3.5 for several probabilities of success.

Table 3.5 The number of electrons N_e for communication between two binary switches for probabilities of success 0.5 and 0.99

L/a	N_e ($\Pi = 0.5$)	N_e ($\Pi = 0.99$)
2	1	7
10	7	44
100	69	459
1000	693	4603
10^4	7×10^3	5×10^4
10^5	7×10^4	5×10^5
10^6	7×10^5	5×10^6
10^7	7×10^6	5×10^7

Since, according to (3.43a) the switching/communication energy per device, E_{sw}, is proportional to the number of electrons, as follows from Table 3.5, communication at distance is energetically a very costly process.

3.4.4 Fan-out costs

In Figure 3.13, an 'upstream' binary switch activates only one 'downstream' device, which can only replicate the input value, i.e. communication. More complex circuits are required for logic operations, where the necessary attributes of logical inference are *convergence* and *branching*. Therefore, for logic operations, the 'upstream' binary switch needs to control at least two other 'downstream' binary switches (Fig. 3.14a). The number of the downstream devices that are driven by a given upstream device is called 'fan-out' (F). The minimum fan-out required for branching is $F_{min} = 2$. In practical systems, fan-out typically ranges between two and four.

The probability that N_e electrons in the interconnect line of device **A** will be found in the gates of *both* **B** and **C** is:

$$\Pi_{B_{AND}C} = \Pi_B \cdot \Pi_C = \Pi_2 = \left(1 - \left(1 - \frac{1}{k}\right)^{N_e}\right)^2 \tag{3.64}$$

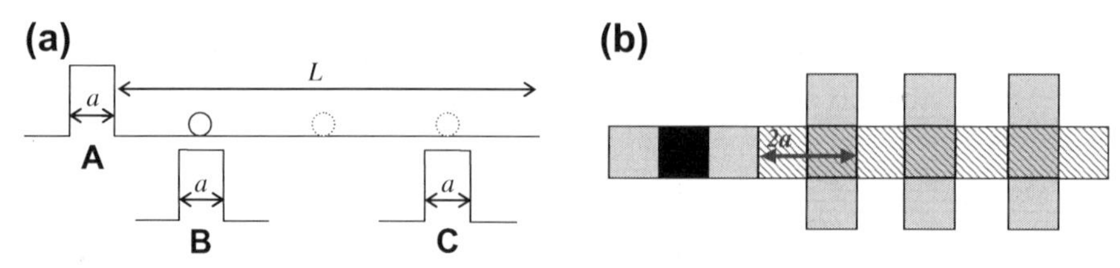

FIGURE 3.14

Device abstraction of connected binary switches: (a) barrier model; (b) generic floorplan

In general

$$\Pi_F = \left(1 - \left(1 - \frac{1}{k}\right)^{N_e}\right)^F \tag{3.65}$$

where F is number of receiving devices, or fan-out. Solving (3.65) for N_e obtain

$$N_e = \frac{\ln\left(1 - \sqrt{\Pi}\right)}{\ln\left(1 - \frac{1}{k}\right)} \tag{3.66}$$

From simple geometrical considerations illustrated in Figure 3.14b, the minimum interconnect length in 2D topology is

$$L_{\min} = 2aF \tag{3.67a}$$

and thus the k term in (3.66), which is the number of interconnect tiles becomes:

$$k_{\min} = \frac{L_{\min}}{a} = \frac{2aF}{a} = 2F \tag{3.67b}$$

Table 3.6 presents the number of electrons needed to guarantee the specified reliability of circuit operation given by (3.66) for the minimum interconnect length.

3.4.5 Energy per tile

In Section 3.3.4, the total switching energy was derived as (3.43a), repeated here for convenience:

$$E_{SW} = (N_e + 2)E_b \tag{3.68a}$$

if $N_e = 1$ and $E_b = k_B T \ln 2$ then

$$E_{SW_{\min}} = 3k_B T \ln 2 \tag{3.68b}$$

Each device (e.g. FET) consists of three tiles and thus the switching energy *per tile* in a device is:

$$\varepsilon_d \approx k_B T \tag{3.68c}$$

One also can calculate the switching energy per *interconnect tile* using (3.66) and (3.68a). Figure 3.15 displays the energy per tile for different interconnect lengths (measured in the number of tiles) for different fan-outs. It can be seen that in the limit of long interconnects, the energy per tile is constant and it varies from about $0.7k_BT$/tile to $1.8k_BT$/tile for $F = 1$–4.

Table 3.6 Fan-out costs

F	N_e ($\Pi = 0.5$)	N_e ($\Pi = 0.99$)
2	5	19
3	9	32
4	14	45
5	20	59
6	26	74

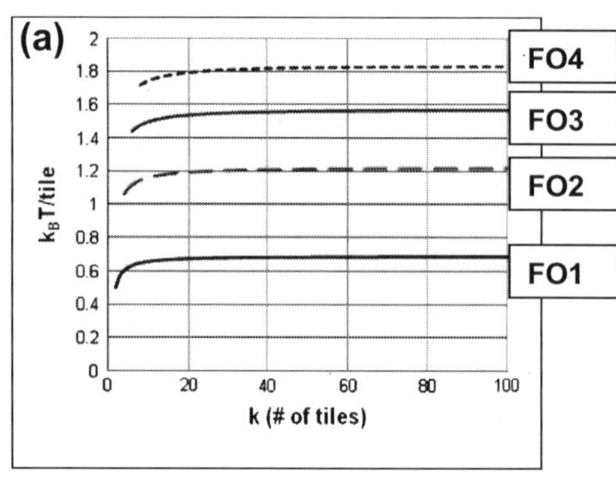

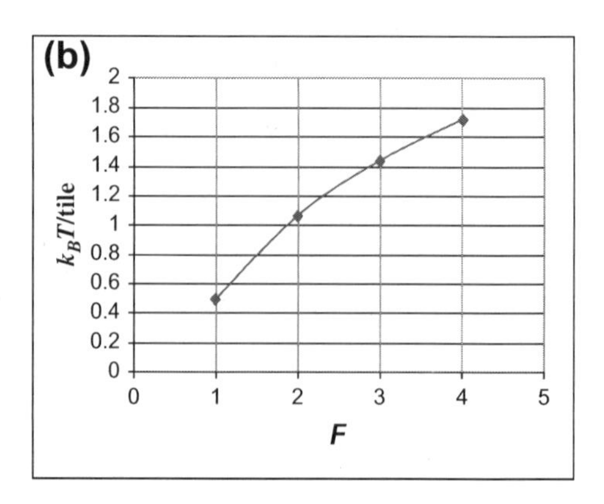

FIGURE 3.15

Interconnect energy per tile: (a) long interconnect limit; (b) minimum interconnect limit

Now, assuming that fanout F is uniformly distributed from $F = 1$ to $F = 4$ between different fragments of a circuit, the average energy per interconnect tile can be calculated from (3.66) and (3.68a) as:

$$\langle \varepsilon_i \rangle_{long} = 1.33 k_B T \ \left(k \rightarrow \infty \right) \tag{3.69a}$$

and

$$\langle \varepsilon_i \rangle_{short} = 1.18 k_B T \ \left(k \rightarrow k_{min} = 2F \right) \tag{3.69b}$$

Comparing (3.68c), (3.69a) and (3.69b), obtains:

$$\varepsilon_d \approx \langle \varepsilon_i \rangle \approx k_B T = \langle \varepsilon \rangle_{tile} \tag{3.69c}$$

i.e. in the limit, the average energy per functional tile of both devices and interconnect is approximately the same.

3.4.6 Implications for nanomorphic cell: Numerical estimates of energy per bit operation

Based on results obtained in previous sections, the energy needed for operation of a micro-sized logic unit that controls operations of a nanomorphic cell can be estimated. In the estimates offered below, a minimum barrier width $a \sim 5$ nm is assumed with corresponding 2D device density, $n \sim 10^{10}$ cm^{-2} (a practical limit projected by the Semiconductor Roadmap [14]). Assume further that the system reliability (the probability of a correct operation of all N switches in a circuit) is $\Pi = 0.99$. The following analysis begins with a larger-scale system, such as a semiconductor chip, to serve as a reference point before considering a small-scale nanomorphic cell.

Large-scale chip: 2D system with size ~1 cm

From (3.54), for $a \sim 5$ nm, a planar lay-out of a digital circuit on the area 1 cm × 1 cm results in $N = 10^{10}$ transistors in the unit. The barrier height required for 99% system reliability is $E_b = 27.63 k_B T$

(from Table 3.4) and the number of electrons per switching event is $N_e = 19$ (from Table 3.6 for a minimum fan-out $F_{min} = 2$). Substituting the data into (3.68a), there results:

$$E_{sw} = (N_e + 2)E_b = (19 + 2) \cdot 27.6k_BT \approx 580k_BT \approx 2 \cdot 10^{-18}J$$

This estimate is consistent with ITRS projections for the switching energy per transistor in microprocessor chips [4].

Small-scale chip: 2D system with size ~10 μm

A planar lay-out of the control unit of the nanomorphic cell on the area 10 μm × 10 μm results in approximately 10^4 transistors in the unit. As a reference, the Intel 8080 microprocessor contained 6000 transistors (Table 3.2). Therefore a 10-μm processor could have reasonable complexity, e.g. sufficient for general purpose computing. The barrier height required for 99% system reliability is $E_b = 13.81k_BT$ (from Table 3.4) and the number of electrons per switching event is $N_e = 19$ (from Table 3. 6 for a minimum fan-out $F = 2$). Substituting the data into (3.68a), we obtain:

$$E_{sw} = (N_e + 2)E_b = (19 + 2) \cdot 13.8k_BT \approx 290k_BT \approx 10^{-18}J$$

As one can see, system scaling allows for approximately two-fold energy reduction per switch.

For a limited energy supply of ~10^{-5} J as estimated in Chapter 2, the total number of binary events that could be performed by the control unit is:

$$N_{bit} = \frac{E_{stored}}{E_{sw}} \sim \frac{10^{-5}J}{10^{-18}J/bit} \sim 10^{13} \text{ binary operations}$$

These estimates suggest that the micro-scale processor could have reasonable complexity and could perform extensive information processing with the available micron-scale energy supply. In the next section further opportunities are discussed that could reduce the energy of logic operations.

3.4.7 Device opportunities for beyond the planar electronic FET: A nanomorphic cell perspective

Opportunities in 3D systems

All of the above derivations were done for planar lay-out, i.e. 2D topology, the basic geometry of mainstream integrated circuits. However a 2D topology also results in long interconnects, and therefore increased energy of operation, since more electrons are required for reliable switching.

In a hypothetical 3D topology for binary switches, the generic shape of the binary switch corresponding to the energy diagram of Figure 3.4 could be a cylinder, and the 3D organization of switches would allow for 'stacked' configurations of Figure 3.11, without as many additional wires as in 2D layout (see Figs 3.12 and 3.14). It could enable 'wireless' communication between the sending devices **A** and several receiving devices, as is shown in Figure 3.16. In principle, as many as six receiving switches can be electrostatically coupled to one sending switch and thus eliminate many circuit connectors. This could dramatically reduce the number of electrons needed for branched communication (fan-out) – since, in principle one electron is sufficient to drive the nearby devices (in practice, several electrons may be needed). Note that the advantages of 3D topology increase with fan-out, thus suggesting a larger fan-out design approach might be desirable for 3D. Even for low fan-out, at least one order of magnitude in energy reduction could be expected (Fig. 3.17).

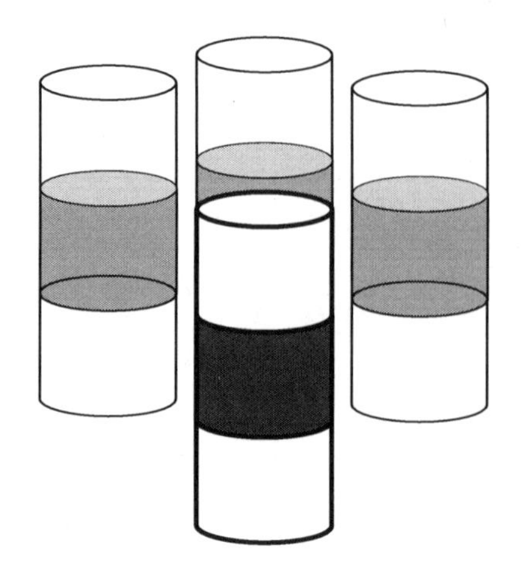

FIGURE 3.16

3D organization of binary switches. The central switch can be electrostatically coupled to up to six other switches (three are shown) without interconnecting wire

Small-scale chip: 3D system with size ~10 μm

In a 3D configuration, the maximum packing density of binary switches is ~10^{17} devices/cm^3 (3.56) which corresponds to ~10^8 devices in a 10-μm cube. From Table 3.4, the corresponding device barrier height is $E_b = 18.79$ eV (99% system reliability). The number of electrons per switching event can be much smaller in 3D than in 2D due to interconnect minimization as discussed above. In fact, in the limit only one electron, $N_e = 1$, is required for fan-out $F = 2–6$. Substituting these data in (3.68a) we obtain:

$$E_{sw} = (N_e + 2)E_b = (1 + 2)\cdot 18.79 k_BT \approx 56 k_BT \approx 2 \times 10^{-19} J$$

Thus, 3D topology could potentially provide significant reductions in energy use, but will require the invention of new devices (e.g. nanowire transistors) and new process integration technologies. Nanowire field effect transistors (NWFET) are currently studied by many groups worldwide [20],

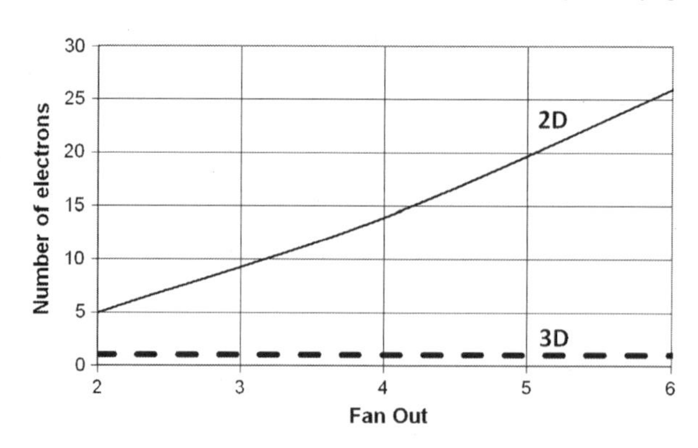

FIGURE 3.17

Fan-out costs in 2D and 3D topologies

and they are regarded as promising candidates both for extremely scaled digital circuits and for other applications such as sensors (as will be discussed in Chapter 4). Studies of scaling limits of nanowire transistors suggest the minimum barrier (channel) length in Si NWFET to be about 5 nm [21].

Devices utilizing information carriers other than electron charge

Nanowire-based FETs (NW FET), while substantially different from planar devices (currently employed in semiconductor chips) are electron-based transistors (though of different geometry), and thus have the same fundamental limits of scaling as planar devices. That is set by the minimum width of the energy barrier, a_{min}. If the barrier is too narrow, excessive quantum mechanical tunneling through the barrier will destroy the operation of transistor. Indeed, tunneling leakage is a major factor which limits transistor scaling. The minimum barrier dimension was derived in Section 3.3.1 and is repeated here for convenience:

$$a \sim \frac{\hbar}{2\sqrt{2mE_b}}, \tag{3.70a}$$

which can be re-written as:

$$a\sqrt{m} \sim \frac{\hbar}{2\sqrt{2E_b}} \tag{3.70b}$$

As was discussed in this chapter, the practical size limit for electron-based devices is ~5 nm.

Note that the size in (3.70) depends inversely on the square-root of mass of information carrier, m, and larger mass would allow for smaller devices, since tunneling would be suppressed. Thus heavier mass information-bearing particles may be a candidate for devices smaller than 5 nm in critical size [22].

The suggestion that a heavier mass information carrier is preferable for nm-scale devices may seem counterintuitive. A common-sense observation is that lighter mass moves faster and requires less energy to move. However, the switching time at a given energy actually remains constant for scaled devices as long as the length-square-root-of-mass product $L\sqrt{m}$ remains constant. An elementary illustration of this statement is shown below, and a more detailed treatment of the problem can be found in [22].

The transfer time needed for a particle of velocity v to move a distance L is:

$$t = \frac{L}{v} \tag{3.71a}$$

Using the relation between particles' velocity and their kinetic energy $E = \frac{mv^2}{2}$, obtain for $L = a$:

$$t = a\sqrt{\frac{m}{2E}} \propto a\sqrt{m} \tag{3.71b}$$

Note also that the tunneling relation (3.70b) contains the product $a\sqrt{m}$, and for fixed energy, the product $a\sqrt{m}$ remains constant for a given tunneling probability. Thus, in principle it might be possible to scale down below electron device limits, if the mass of the information-bearing particle is increased properly, such as:

$$a\sqrt{m} = const \tag{3.72a}$$

or

$$m_{opt} \sim \frac{1}{a^2} \tag{3.72b}$$

Note that for dimensional scaling with barrier height/operational energy held constant, the switching time will remain constant, i.e. devices using heavier particles should not be inferior to electron-based devices.

Several recent demonstrations indeed suggest the possibility of physical realizations for a sub-5 nm binary switch. The atomic-scale switch reported in [5, 23], for example, opens or closes an electrical circuit by the controlled reconfiguration of silver atoms within an atomic-scale junction. Such 'atomic relays' operate at room temperature and the only moveable parts of the switch are the contacting atoms, which open and close a nm-scale gap. Experimentally, a critical device size (gap) of 1 nm was reported [23]. The atomic relay operates at a relatively low voltage of 0.6 V. The experimentally measured switching time was 1 μs, though the authors projected the switching time for optimized devices will be in the range of 1 ns [23].

Moving atoms/ions also plays a key role in the mechanism for the operation of a recently reported 'memristor', utilizing TiO_2 thin films, where the switching occurs due to ionic motion of oxygen vacancies [8, 24, 25]. Memristor-type devices may have a potential for extreme scaling. Also, some researchers believe that memristors could offer a new fundamental circuit element, which could allow for realization of complex functions with lower device count. The latter is very important for nanomorphic cell applications, where volume is one of the primary concerns. The model of memristive behavior has recently been proposed as a possible mechanism in the adaptive behavior of unicellular organisms (amoebas) [25], and it was suggested that memristors could be used to build biology-inspired electronic circuits [25].

As a final remark, ions in liquid electrolytes play an important role in biological information processors, such as the brain. For example, in the human brain, the distribution of calcium ions in dendrites may represent another crucial variable for processing and storing information [9]. Calcium ions enter the dendrites through voltage-gated channels in a membrane, and this leads to rapid local modulations of calcium concentration within dendritic tree [9]. Based on the brain analogy, the binary state could be realized by a single ion that can be moved to one of two defined positions, separated by a membrane (the barrier) with voltage-controlled conductance. These or similar structures might be used to make an atom-based binary switch scalable to ~1 nm or below.

3.5 SUMMARY

In this chapter bounds for energy and complexity of a nanomorphic implementation of logic control unit were developed based on fundamental physics and assuming ideal conditions/best case scenarios. The idea of an energy barrier was developed and used as a unifying concept for the analysis of the physical scaling limits for the binary switch (e.g. field effect transistor) and for estimates of switching energies and times. Also the interconnect lines were included using a model of the probability of transmitted electron location that was ultimately used to quantify energy losses. Results of the derivations in this chapter are summarized in Table 3.7. It was shown that the micro-scale processor could have reasonable complexity and perform extensive information processing with the available micron-scale energy supply discussed in Chapter 2.

Aggressive FET scaling is mandatory for implementation of micron-scale systems, in order to achieve a full functionality of the logic unit. Transition from planar (2D) layout to a 3D configuration could result in a significant energy reduction. A summary of projected characteristic device critical

Table 3.7 Projected device characteristics for binary switches for different system size and topology

	System size	System Topology	E_{bit}	a, nm	n	N
Lower bound		2D, 3D	$\sim 3k_BT = 10^{-20}$ J	1 nm	$\sim 10^{13}$ cm^{-2} $\sim 10^{19}$ cm^{-3}	–
2022 planar FET	~ cm	2D	$600k_BT \sim 2 \cdot 10^{-18}$ J	5 nm	10^{10} cm^{-2}	10^{10}
2022 planar FET	~10 μm	2D	$260k_BT \sim 10^{-18}$ J	5 nm	10^{10} cm^{-2}	10^4
2022 NW FET	~10 μm	3D	$56k_BT \sim 2 \cdot 10^{-19}$ J	5 nm	10^{17} cm^{-3}	10^7

size (a), operation energy (E_{bit}), device density (n) and the total number of devices in system (N) is given in Table 3.7. The transition to 3D circuits will require a new device geometry, e.g. cylindrical *nanowire*, and new interconnect strategies.

Devices beyond FET also need to be analyzed for their potential to offer more functionality at less device count and/or smaller energy of operation. In the *nanomorphic cell*, volume utilization and energy minimization are the two main criteria for device evaluation.

The volumetric constraints imposed by the nanomorphic cell suggest device scaling beyond the limits of the electron-charge-based FET needs to be explored. Device scaling below 5 nm might utilize information-bearing particles whose mass exceeds that of the electron. It has been shown theoretically that 'atomic switches' based on moving atoms as information carriers can offer superb switching characteristics relative to electron-based devices in deep nanometer domain [22]. 'Nanoionic' devices have already attracted the attention of several research groups and several promising experimental devices have been recently demonstrated [5,6,8,23–25].

APPENDIX 1: QUANTUM CONFINEMENT

Consider a particle (e.g. an electron) confined in a one-dimensional potential well of width w with abrupt (vertical) walls (Fig. A1.1). The Heisenberg coordinate-momentum relation (3.5b) for $\alpha = 1$ and $\Delta x = w$ results in

$$\Delta p \cdot w \geq h \tag{A1.1}$$

Let the electron possess a certain kinetic energy E and therefore, a momentum $\pm p = \sqrt{2mE}$ (the $\pm$ reflects the fact that the direction of the momentum is undefined (i.e. right or left). Thus

$$\Delta p = p - \left(-p\right) = 2p = 2\sqrt{2mE} \tag{A1.2}$$

Substituting (A1.2) in (A1.1) results in

$$2w\sqrt{2mE} \geq h \tag{A1.3}$$

or

$$w\sqrt{2mE} \geq \frac{h}{2} \tag{A1.4}$$

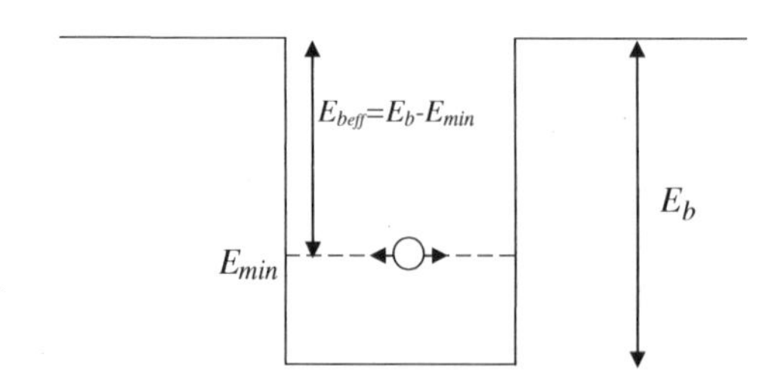

FIGURE A1.1

Electron confined in a rectangular well

Thus the minimum kinetic energy (also called the ground-state energy) of a particle in a well is

$$E_{\min} = \frac{h^2}{8mw^2} \tag{A1.5}$$

Equation (A1.5) coincides with a standard solution of a quantum mechanical problem for a particle in a rectangular box with infinite walls [11].

Note that the ground state energy E_{min} is always above the bottom of the well and it moves higher when the width of the well decreases, an effect called *quantum confinement*. Quantum confinement sets a limit on the minimum width of the well for the binary switch shown in Figure 3.4. If the well is formed by barriers of finite height E_b, the effective barrier height for a particle confined in the well is less than E_b:

$$E_{b_{eff}} = E_b - E_{\min} \tag{A1.6}$$

If the effective barrier becomes very small, the particle can jump over due to, e.g., thermal excitations. Thus an estimate for the minimum size of the well is suggested by $E_b = E_{min}$. From (A1.5):

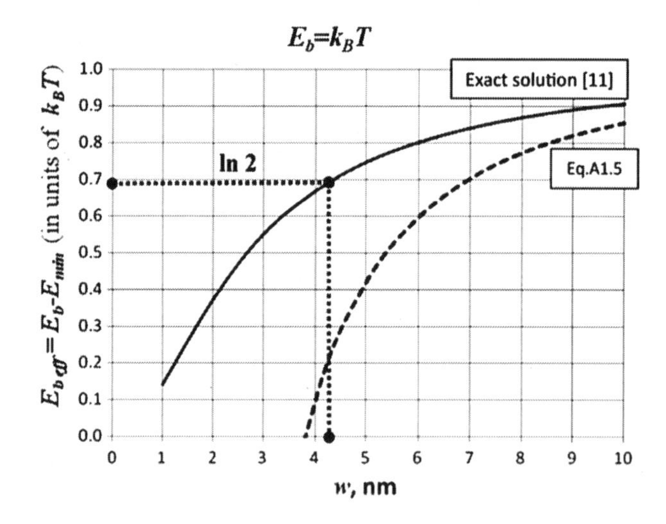

FIGURE A1.2

Effective barrier height for an electron confined in a potential well

$$w_{\min} \sim \frac{h}{2\sqrt{2mE_b}} \tag{A1.7}$$

Note that (A1.7) is close to the *Heisenberg distinguishability length* a_H (3.10) derived for tunneling.

Generally speaking, Eq. (A1.5) describes a situation of infinite height of the walls [11], and for a finite barrier height, the solution is obtained numerically (see e.g. [11] for a detailed procedure). However (A1.5) can still be used for order-of-magnitude estimates in the case of low barriers. For example, let $E_b = k_BT$, then the effective barrier for a confined particle as a function of the well size is plotted in Figure A1.2 calculated using approximation (A1.5) and by an exact numerical solution [11]. As can be seen, both approaches yield similar result for larger w, e.g. $w \sim 10$ nm and diverge for smaller w, remaining however within a reasonable range for order-of-magnitude estimates. As an interesting observation, the simple approximation (Eqs. A1.5 and A1.7) suggests $w_{min} \sim 4$ nm, where the effective barrier for electron becomes very small. On the other hand, the exact solution says that at $w \sim 4$ nm the effective barrier height is $\sim k_BT\ln 2$, i.e. the Boltzmann's limit for the minimum barrier height (3.18).

APPENDIX 2: DERIVATION OF ELECTRON TRAVEL TIME (EQ. 3.55)

The travel time of the electron along the distance L is determined by electron's average velocity $\langle v \rangle$:

$$\tau \sim \frac{L}{\langle v \rangle} \tag{A2.1}$$

In this treatment, ballistic transport is assumed as the best-case scenario, and thus constant acceleration motion, for which

$$\langle v \rangle = \frac{v_{\min} + v_{\max}}{2} \approx \frac{v_{\max}}{2} \tag{A2.2}$$

(v_{min} is assumed to be zero).

The maximum velocity v_{max} can be found from the energy balance relation:

$$E = \frac{mv_{\max}^2}{2} \tag{A2.3}$$

From (A2.2) and (A2.3):

$$\langle v \rangle = \frac{v_{\max}}{2} = \sqrt{\frac{E}{2m}} \tag{A2.4}$$

and from (A2.1) and(A2.4) the electron's travel time is:

$$\tau \sim L\sqrt{\frac{2m}{E}} \tag{A2.5}$$

Next, from Table 3.3:

$$\frac{\tau_H}{a_H} = \frac{\hbar}{2E} \cdot \frac{2\sqrt{2mE}}{\hbar} = \frac{\sqrt{2mE}}{E} = \sqrt{\frac{2m}{E}} \tag{A2.6}$$

Substituting (A2.6) into (A2.5) we obtain the result for minimum travel time (3.55):

$$\tau = \frac{L}{a_H}\tau_H \tag{A2.7}$$

LIST OF SYMBOLS

Symbol	Meaning
a	Energy barrier width, tile size
a_H	Heisenberg distinguishability length
a_{HB}	Boltzmann-Heisenberg length
BIT	Maximum binary throughput
C	Capacitance
d	Distance
e	Electron charge, $e = 1.6 \times 10^{-19}$ C
E	Energy
E_b	Energy barrier height
E_{sw}	Switching energy
f	Frequency
$\mathbf{F}$	Force
F	Fan-out
G	Conductance
G_0	Quantum conductance, $G_0 = 7.75 \times 10^{-5}$ A/V
h	Planck's constant, $h = 6.63 \times 10^{-34}$ J·s
$\hbar$	Reduced Planck's constant $\hbar = h/2p = 1.055 \times 10^{-34}$ J·s
I	Current
k	Number of tiles
k_B	Boltzmann constant, $k_B = 1.38 \times 10^{-23}$ J/K
L, l	Length
m	Mass
m_e	Electron mass, $m_e = 9.31 \times 10^{-31}$ kg
n	Device packing density
n_i	Intrinsic carrier concentration in an undoped semiconductor (for Si, $n_i = 1.45 \times 10^{10}$ cm^{-3})
N	Number of binary switches
N_a^-	Concentration of ionized acceptor impurities in semiconductor
N_d^+	Concentration of ionized donor impurities in semiconductor
N_e	Number of electrons
p	Momentum
P	Power
R	Resistance
R_0	Quantum resistance, $R_0 = 12.9$ kΩ

Symbol	Meaning
S	Action
t, τ	Time, time interval
t_H	Heisenberg time
T	Absolute temperature
v	Velocity
V	Voltage
V_g	Gate voltage
V_t	Threshold voltage
w	Width of potential well
x	Coordinate
α	Constant
ε_0	Permittivity of free space, $\varepsilon_0 = 8.85 \cdot 10^{-12}$ F/m
ε_d	Energy per device tile
$\langle \varepsilon_i \rangle$	Average energy per interconnect tile
$\langle \varepsilon \rangle_{tile}$	Average energy per tile
Π	Probability
ρ	Charge density
τ_H	Heisenberg time
φ	Electric potential
$\sim$	Indicates order of magnitude

References

[1] R.U. Ayres, Information, Entropy and Progress, AIP Press, New York, 1994.
[2] L. Brillouin, Science and Information Theory, Academic Press, New York, 1962.
[3] R. Waser (Ed.), Nanoelectronics and Information Technology, Wiley-VCH, Weinheim, 2003.
[4] V.V. Zhirnov, R.K. Cavin, Physics of Computational Elements, in: R. Waser (Ed.), Nanotechnology.Volume3: Information Technology I., WILEY-VCH Verlag GmbH & Co, KGaA, Weinheim, 2008.
[5] K. Terabe, T. Hasegawa, C. Liang, M. Aono, Control of local ion transport to create unique functional nanodevices based on ionic conductors, Sci. Technol. Adv. Mater. 8 (2007) 536–542.
[6] R. Waser, M. Aono, Nanoionics-based resistive switching memories, Nature Mater 6 (2007) 833–840.
[7] V.V. Zhirnov, R.K. Cavin, Charge of the heavy brigade, Nature Nanotech. 3 (2008) 377–378.
[8] D.B. Strukov, J.L. Borghetti, R.S. Williams, Coupled ionic and electronic transport model of thin-film semiconductor memristive behavior, SMALL 5 (2009) 1058–1063.
[9] C. Koch, Computation and a single neuron, Nature 385 (1997) 207–210.
[10] R. Gomer, Field Emission and Field Ionization, s.l. : Harvard University Press, 1961.
[11] C. Cohen-Tannoudji, B. Diu, F. Laloë, Quantum Mechanics, Hermann and John Wiley & Sons, 1977.
[12] N. Margolus, L.B. Levitin, The maximum speed of dynamical evolution, *Physica* D 120 (1998) 1881.
[13] D.C. Brody, Elementary derivation for passage times, J. Phys. A-Math and General 36 (2003) 5587.

[14] The International Technology Roadmap for Semiconductors (2007). <http://www.itrs.net/>.
[15] S.M. Sze, Physics of Semiconductor Devices, John Wiley & Sons, 1981.
[16] R.K. Cavin, V.V. Zhirnov, J.A. Hutchby, G.I. Bourianoff, Energy barriers, demons, and minimum energy operation of electronic devices, Fluctuation and Noise Lett. 5 (2005) C29.
[17] Y. Taur, T.H. Ning, Fundamentals of Modern VLSI Devices, Cambridge University Press, 1998.
[18] J.W. Joyner, Limits on device packing density as 2-D tiling problem, *unpublished*
[19] S. Shankar, V. Zhirnov, R. Cavin, Computation from devices to system level thermodynamics, ECS Trans. 25 (7) (2009) 421–431.
[20] W. Lu, P. Xie, C.M. Lieber, Nanowire transistor performance limits and applications, IEEE Trans. Electron Dev. 55 (2008) 2859.
[21] B. Yu, L. Wang, Y. Yuan, P.M. Asbeck, Y. Taur, Scaling of nanowire transistors, IEEE Trans. Electron Dev. 55 (2008) 2846.
[22] V.V. Zhirnov, R.K. Cavin, Emerging research nanoelectronic devices: the choice of information carrier, ECS Trans. 11 (6) (2007) 17–28.
[23] K. Terabe, T. Hasegawa, T. Nakayama, M. Aono, Quantized conductance atomic switch, Nature 433 (2005) 47–50.
[24] D.B. Strukov, G.S. Snider, D.R. Stewart, R.S. Williams, The missing memristor found, Nature 453 (2008) 80–83.
[25] M. Di Ventra, Y.V. Pershin, L.O. Chua, Circuit elements with memory: memristors, memcapacitors and meminductors, Proc. IEEE 97 (2009) 1717.
[26] J.B. Keller, Corrected Bohr-Sommerfeld quantum conditions for nonseparable systems, Ann. Phys. 4 (1958) 180–188.
[27] I.P. Batra, Origin of conductance quantization, Surf. Sci. 395 (1998) 43–45.
[28] E. Scheer, N. Agrait, J.C. Cuevas, A.L. Yeyati, B. Ludoph, A. Martin-Rodero, et al., The signature of chemical valence in the electrical conduction through a single-atom contact, Nature 394 (1998) 154–157.
[29] Y. Imry, R. Landauer, Conductance viewed as transmission, Rev. Mod. Phys. 71 (1999) S306–S312.

CHAPTER

Sensors at the micro-scale

4

CHAPTER OUTLINE

4.1 INTRODUCTION

The nanomorphic cell must have mechanisms that enable it to obtain information about the external environment and, in particular, about the state of a target living cell. It is stipulated that the nanomorphic cell must measure external parameters in a non-invasive way. Thus, an essential function of the nanomorphic cell for bio-applications is to receive and transform signals from its environment and biological signals from the living cell into an electrical form for subsequent processing and analysis to provide a basis for further actions. For the purposes of this discussion, the sensors must be of a size commensurate with that of the artificial cell.

In this chapter, the basic principles of sensor elements are discussed and scaling projections for micron-sized sensors are investigated. The barrier model (as in the FET) is shown to be a generic abstraction for many classes of sensors. The concept of the signal-to-noise ratio is important for understanding sensor scaling while the barrier formulation is used to derive a model for thermal Johnson-Nyquist noise. It is argued that in many cases, sensors can be regarded as either *digital* or *analog* devices (e.g. transistors) and that the scaling limits of sensors are therefore close to the limits of the binary switches, which were derived in Chapter 3 from basic physics. In fact, the field-effect transistor is used in many practical implementations of chemical sensors. It is shown that a nanowire-based FET performs the best for many bio-chemical sensing applications.

The discussion in this chapter is focused on three types of sensors: (i) bio-electrical; (ii) bio-chemical; and (iii) bio-thermal. While other types of sensors, for example light or pressure sensors, are also essential for nanomorphic measurements, the size and scope of this book does not allow for an inclusive study of different sensors. In principle, scaling behavior of other types of sensors can also be analyzed within the barrier model framework and readers may find such an exercise worthwhile.

This chapter, by its nature, is an integration of concepts from both biological and physical sciences. To aid the reader, a brief glossary of biological terminology used herein is included at the end of the chapter.

4.2 SENSOR BASICS

A sensor is a device which converts an external physical stimulus into a distinguishable and processable signal, usually in electrical form [1]. Sensors are sometimes defined as a whole system, comprising, e.g., (i) a transducer, which generates an electrical signal in response to an external stimulus, (ii) an electronic amplifier, which increases the intensity of the signal, (iii) a signal processor, (iv) a display, etc. [1]. In this text, a narrower definition of a sensor is used: a device directly responding to external stimuli, i.e. a transducer.

Examples of external physical stimuli are *mechanical* (e.g. pressure, motion, vibration), *electrical* (e.g. voltage), *thermal* (e.g. temperature difference), *electromagnetic* (e.g. light), *chemical* (e.g. presence/absence of particular chemical species), etc.

In response to the external stimulus, the sensor generates an electrical signal, which is further processed by a control unit and eventually provides a basis for further actions by the nanomorphic cell. In its simplest form, a sensor generates a binary YES/NO response by distinguishing between the presence and absence of a particular external stimulus. In other words, a sensor must have at least two distinguishable states. As was discussed in Chapter 3, creation of the distinguishable states requires energy barriers within the sensing device. Indeed, in many cases, a sensor can be regarded as a switch, whose barrier is deformed by different stimuli, e.g. pressure, light, temperature, presence of ions in a solution, etc. Note that the FET discussed in Chapter 3 is indeed an *electrical sensor*, whose stimulus is the presence/absence of an electrical charge on the gate electrode. In fact, a charge-sensing transistor, e.g. the FET, is usually a receiving device downstream from the sensor that is used for signal conditioning (e.g. amplification, filtering, etc.), as is shown in Figure 4.1.

Due to the intrinsic similarity between a sensor and the FET, the fundamental scaling limits of sensors are, in principle, the same for the FET, which were discussed in Chapter 3. However, there are additional constraints for sensors arising from requirements of *sensitivity*, *selectivity*, and *response time*, as will be discussed in this chapter. Also, while sensors can be used in a digital (i.e. binary) mode, in most cases they are used as *analog* devices (see Box 4.1). As a side note, some sensors can be powered, at least partially, by the energy of the external stimulus. For example, a thermoelectric temperature sensor (e.g. thermocouple/thermopile) can be used as an energy source converting heat into electricity, as was discussed in Chapter 2.

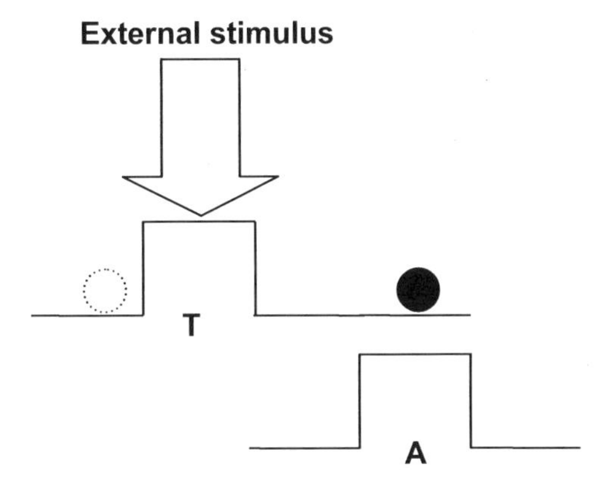

FIGURE 4.1

A generic model of a minimal sensor configuration: transducer (T) and a downstream device used for signal conditioning, e.g. amplifier (A)

BOX 4.1 DIGITAL AND ANALOG DEVICES

The binary switches, discussed in Chapter 3, are used to represent and process information in *digital* form: they have two states (known as binary states), which are marked as the digits **0** and **1** (see Box 3.1). As was discussed in Chapter 3, in digital operation, the barriers in the switches (e.g. the FET) are abruptly set from *high* to *low* values and vice versa.

The barrier height can also be decreased gradually to a value somewhere between its maximum and minimum values. Devices (e.g. the FET) operating in the continuous mode are called *analog*, while devices operating in the discrete (e.g. binary) mode are called *digital*. Most sensors are *analog* devices.

In many microelectronic systems, both analog and digital parts co-exist. To enable interaction between the two, signal converters are used: *analog-to-digital* (ADC) and *digital-to-analog* converters (DAC).

4.3 ANALOG SIGNAL

Figure 4.2 contrasts analog and digital forms of data representation. An analog signal is a time-varying quantity $s(t)$, which if transformed into digital form (e.g. by an analog-to-digital converter ADC), each point of $s(t)$ is then represented by a sequence of pulses of constant duration and amplitude. Note that measurement of a continuous analog quantity requires a certain finite time interval, τ, and the observed value of $s(t)$ is a *time average* over the interval τ:

$$\bar{s}_\tau = \frac{1}{\tau}\int_0^\tau s(t)dt \tag{4.1}$$

Obviously, the recorded time average value $\bar{s}_\tau$ will depend on the measurement time τ and on how rapidly the analog signal changes in time, which can be characterized by the signal's highest and lowest frequencies, f_H and f_L respectively. The difference $\Delta f = f_H - f_L$ is called bandwidth. In the

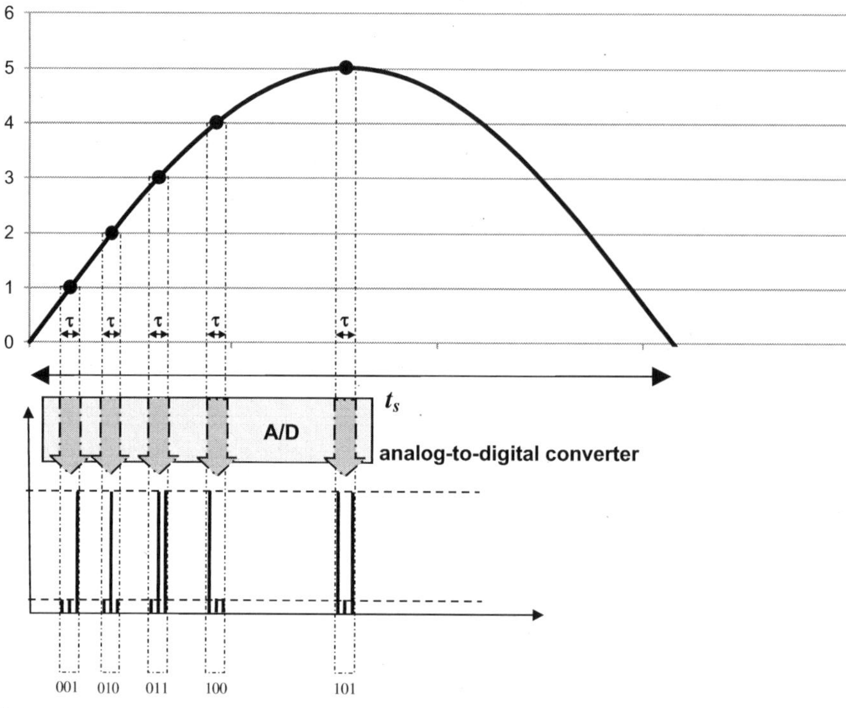

FIGURE 4.2

Examples of *analog* (top) and corresponding *digital* (bottom) signals. The digital signals shown in the figure represent several sample points of the analog signal, generated by an analog-to-digital converter. Recording/conversion of each point of the analog signal requires certain sampling time τ which should typically be much smaller than the signal duration t_s

discussion below it is always assumed that $f_L = 0$ (Hz). The highest frequency (and therefore the bandwidth) is related to the signal's time duration, t_s, as

$$\Delta f = f_H \sim \frac{1}{t_s} \tag{4.2}$$

Note that in theory, for a signal of finite time duration, $f_H \to \infty$. In practice however, the highest frequency of a signal is understood to be that frequency below which the main (e.g. 90%) part of the signal's energy is located. For such a 'relaxed' definition, formula (4.2) holds for most practical cases.

As can be seen from Figure 4.2, for reliable analog recording, the recording interval, τ, (sampling time) must be much less than the signal duration: $\tau << t_s$ or $f_s >> f_H$.

The fundamental limit on the minimum sampling time is given by the Nyquist-Shannon-Kolmogorov theorem:

$$\tau_{\min} \leq \frac{1}{2\Delta f} \sim \frac{t_s}{2} \tag{4.3}$$

To illustrate (4.3), consider strategies for detecting a single rectangular pulse of duration t_s (Fig. 4.3) by using a repetition of rectangular sampling pulses of duration τ. Two things need to happen for a successful detection: (i) the sampling signal needs to 'catch' the measured signal, i.e. there must be a time overlap between the two, and (ii) the recorded time average value $\bar{s}_\tau$ of the detected signal should be reasonably close to the actual value, s_m. In the following, we require that in a 'minimal successful measurement,' the measured value should be no less than a half of the actual value, i.e. $\bar{s}_\tau \geq \frac{s_m}{2}$. First let $\tau >> t_s$ (Fig. 4.3b). In this case, the probability the sampling and the measured signals overlap is close to 1. However, the measured time average value of the signal over interval τ, according to (4.1) is $\bar{s}_\tau \rightarrow 0$. For more accurate time average measurements, τ should be reduced, e.g. $\tau \sim t_s$ (Fig. 4.3c). However, if the repetition interval is larger than t_s, there is a non-zero probability that the sampling will miss the measured signal. If $\tau < t_s$, the overlap probability is always 1 and it is easy to show that if $\tau = t_s/2$, then $\bar{s}_\tau \geq s_m/2$ (Fig. 4.3d).

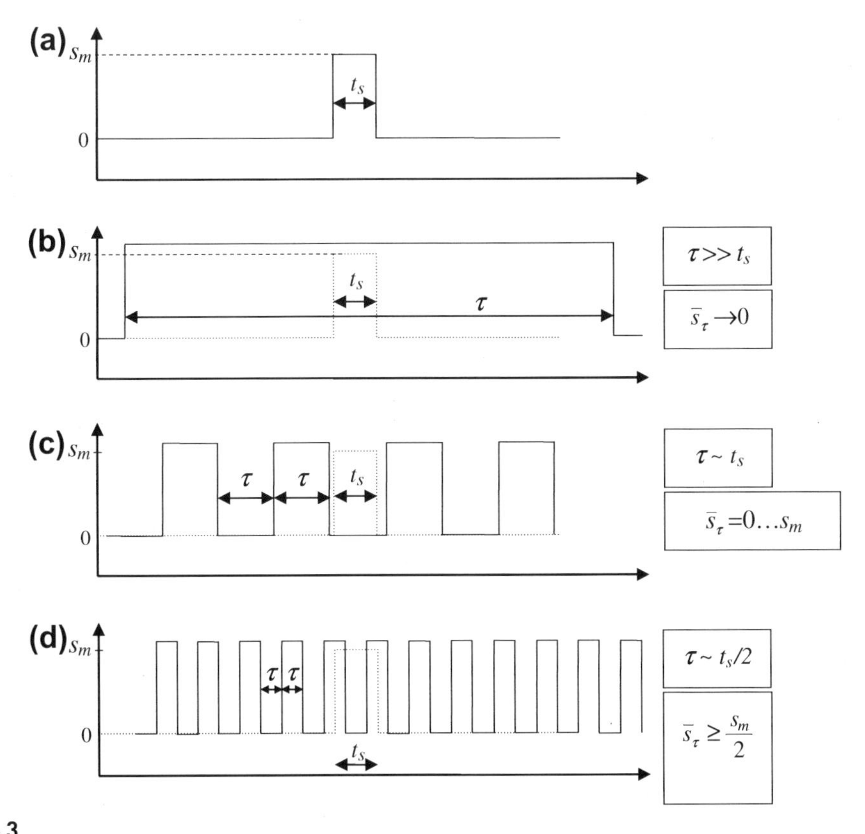

FIGURE 4.3

Illustration to the sampling theorem: different strategies for detection of a single rectangular pulse (see text for discussion)

4.4 FUNDAMENTAL SENSITIVITY LIMIT OF SENSORS: THERMAL NOISE

In Chapter 3, a detailed analysis of the barrier structure (Fig. 3.5) for digital operation (binary switch) was performed and a basic FET equation (3.41) was derived. In fact, the FET equation (3.41) also describes the gradual or analog operation when the controlling gate voltage V_g varies continuously, resulting in continuous change in the current I, passing through the FET:

$$I = eN_0 \exp\left(-\frac{E_{b_0} - eV_g}{k_B T}\right)\left[1 - \exp\left(-\frac{eV_{AB}}{k_B T}\right)\right] \tag{4.4}$$

As was shown in Chapter 3, the limiting factor for operation of digital binary switches is *error probability*. In analog devices, the equivalent measure of errors that limit the ability to detect a small signal is *noise*.

In an ideal digital switch, the current (4.4) is very low (nearly zero) in a steady state and it abruptly changes to a high value for a very short switching event. However, in the steady state of analog applications, there is a non-zero current I_0. When an external signal V_g is applied to the gate, the current changes to a value I_s. One measure of *sensitivity* of a sensor is an ability to distinguish between I_s and I_0. A 'minimal' requirement on the ratio of the signal current to the steady-state current can be written as:

$$\frac{I_s}{I_0} = 2 \tag{4.5a}$$

or for output signal amplitude $\Delta I = I_s - I_0$:

$$\frac{\Delta I}{I_0} = \frac{I_s - I_0}{I_0} = \frac{I_s}{I_0} - 1 = 1 \tag{4.5b}$$

From (4.4) and (4.5), a minimal input signal amplitude $\Delta V_s = V_{gs} - V_{g0}$ can be calculated as

$$\Delta V_s = \frac{k_B T \ln 2}{e} = 18mV \tag{4.6}$$

Note that (4.6) is equivalent to (3.18).

Next, the derivation of (3.41), now (4.4), is revisited, keeping in mind the analog operation of the same structure (Fig. 4.4), and an expression for thermal noise is derived. This derivation was inspired by the elegant treatment by Luca Callegaro at the Istituto Nazionale di Ricerca Metrologica (Italy) [2].

The *average current* flowing through the barrier structure during a *finite* sampling time τ will be first derived. If only one electron passes the barrier during the sampling time, the effective recorded current is

$$i = \frac{e}{\tau} \tag{4.7}$$

The probability that at least one electron will pass the barrier during the time τ can be determined as follows. The probabilities that an electron striking a barrier will pass over the barrier is given by the Boltzmann probability discussed in Chapter 3 (e.g. 3.4):

$$\Pi_{AB} = \exp\left(-\frac{E_b}{k_B T}\right) \tag{4.8a}$$

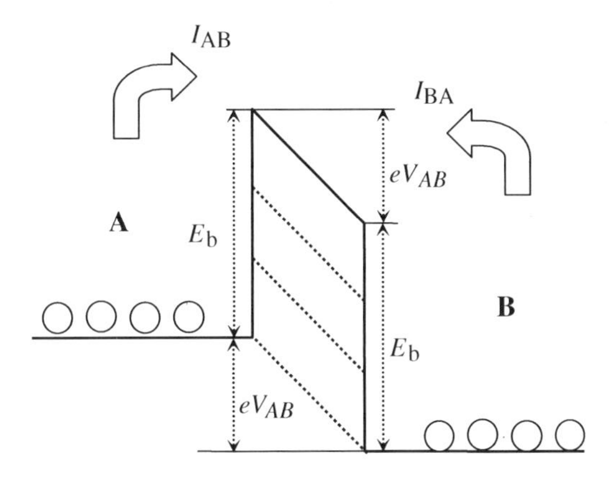

FIGURE 4.4

A barrier structure depicting analog operation (dashed lines show intermediate reduction in barrier height, E_b)

$$\Pi_{BA} = \exp\left(-\frac{E_b + eV_{AB}}{k_B T}\right) \tag{4.8b}$$

Note that:

$$\frac{\Pi_{BA}}{\Pi_{AB}} = \exp\left(-\frac{eV_{AB}}{k_B T}\right) \tag{4.8c}$$

Π_{AB} and Π_{BA} are the probabilities of 'success' in one trial. The number of trials k during time interval τ is

$$k = \tau \cdot N_0 \tag{4.9}$$

where N_0 is the number of electrons that strike the barrier per unit time (in probability language, N_0 is the attempt frequency).

The probability that at least one electron will pass through the barrier during a sampling time τ is

$$\pi(k) = 1 - (1 - \Pi)^k \tag{4.10a}$$

For smaller k:

$$\pi(k) \approx k \cdot \Pi \tag{4.10b}$$

Now, the average value of the barrier current and the mean square value can be calculated using standard formulae for weighted mean:

$$\langle I \rangle = \sum_i \pi_i I_i \tag{4.11a}$$

$$\langle I^2 \rangle = \sum_i \pi_i I_i^2 \tag{4.11b}$$

$$\sum_i \pi_i = 1 \tag{4.11c}$$

The electrons strike the barrier from both sides and there are three possible random events contributing to the current.

1. One electron passes the barrier from **A** to **B**, with probability π_{AB} resulting in a current i_{AB}:

$$\pi_{AB}(\tau) \approx k \cdot \Pi_{AB} = \tau \cdot N_0 \exp\left(-\frac{E_b}{kT_B}\right) \tag{4.12a}$$

$$i_{AB} = \frac{e}{\tau} \tag{4.12b}$$

2. One electron passes the barrier from **B** to **A**, with probability π_{BA} resulting in a current i_{BA}:

$$\pi_{BA}(t) = k \cdot \Pi_{BA} = \tau \cdot N_0 \exp\left(-\frac{E_b + eV_{AB}}{kT_B}\right) \tag{4.12c}$$

$$i_{BA} = \frac{e}{\tau} \tag{4.12d}$$

3. No electrons pass the barrier in either direction during the sampling time τ with probability π_0:

$$\pi_0 = 1 - (\pi_{AB} + \pi_{BA}) \tag{4.12e}$$

$$i_0 = 0 \tag{4.12f}$$

Using (4.11a) and (4.12 a–f), the average current is:

$$\begin{aligned}\langle I\rangle &= i_{AB} \cdot \pi_{AB}(t) - i_{BA} \cdot \pi_{BA}(t) + 0 \cdot (1 - (\pi_{AB} + \pi_{BA}) = \frac{e}{\tau} \cdot \tau N_0 (\Pi_{AB} - \Pi_{BA}) \\ &= eN_0(\Pi_{AB} - \Pi_{BA})\end{aligned} \tag{4.13a}$$

Note that the average current is independent of the sampling time. One can re-write (4.13a) as

$$\langle I\rangle = eN_0(\Pi_{AB} - \Pi_{BA}) = eN_0 \exp\left(-\frac{E_b}{k_B T}\right)\left(1 - \exp\left(-\frac{eV_{AB}}{k_B T}\right)\right) \tag{4.13b}$$

which is identical to (3.41a).

The current can also be expressed using Ohm's law: $I = V_{AB}/R$, where R is an effective resistance of the barrier structure:

$$\langle I\rangle = \frac{V_{AB}}{R} = eN_0\Pi_{AB}\left(1 - \exp\left(-\frac{eV_{AB}}{k_B T}\right)\right) \tag{4.13c}$$

from which using (4.8c):

$$\Pi_{AB} = \frac{V_{AB}}{R}\frac{1}{eN_0}\left(\frac{1}{1-\exp\left(-\frac{eV_{AB}}{k_B T}\right)}\right) \tag{4.14}$$

The mean square value of the current according to (4.11a) and (4.12 a–f) is

$$\langle I^2\rangle = i_A^2 \cdot \pi_{AB}(\tau) + i_B^2 \cdot \pi_{BA}(\tau) = \frac{e^2}{\tau} N_0 (\Pi_{AB} + \Pi_{BA}) \tag{4.15a}$$

or using (4.8c):

$$\langle I^2 \rangle = \frac{e^2}{\tau} N_0 \left(\Pi_{AB} + \frac{\Pi_{BA}}{\Pi_{AB}} \Pi_{AB} \right) = \frac{e^2}{\tau} N_0 \Pi_{AB} \left(1 + \frac{\Pi_{BA}}{\Pi_{AB}} \right) = \frac{e^2}{\tau} N_0 \cdot \Pi_{AB} \left(1 + \exp\left(-\frac{eV_{AB}}{k_B T} \right) \right) \tag{4.15b}$$

Substituting (4.14) for Π_{AB} we obtain:

$$\langle I^2 \rangle = \frac{e}{\tau} \frac{V_{AB}}{R} \cdot \frac{1 + \exp\left(-\frac{eV_{AB}}{k_B T} \right)}{1 - \exp\left(-\frac{eV_{AB}}{k_B T} \right)} \tag{4.15c}$$

Let $x = eV_{AB}/k_B T$ and investigate the case when $V_{AB} \to 0$. For small x, $\exp(-x) \sim 1 - x$, and thus

$$\langle I^2 \rangle = \frac{e}{\tau} \cdot \frac{V_{AB}}{R} \cdot \frac{1 + (1 - x)}{1 - (1 - x)} = \frac{e}{\tau} \frac{V_{AB}}{R} \frac{2 - x}{x} \approx \frac{e}{\tau} \frac{V_{AB}}{R} \frac{2}{x} = 2 \frac{e}{\tau} \frac{V_{AB}}{R} \frac{k_B T}{eV_{AB}} = \frac{2k_B T}{R} \frac{1}{\tau} \tag{4.16}$$

Finally, using the fundamental relation between the signal bandwidth Δf and the minimum sampling time (4.3) we obtain:

$$\langle I_n^2 \rangle = \frac{4k_B T}{R} \Delta f \tag{4.17a}$$

or

$$\langle V_n^2 \rangle = \langle I_n^2 \rangle \cdot R^2 = 4k_B T R \Delta f \tag{4.17b}$$

The subscript on the left-hand side of (4.17a–b) stands for ‘noise’, as the above results indicate that non-zero values of electrical signal can be measured even in the absence of external bias, i.e. $V_{AB} \to 0$. Indeed (4.17a–b) are the Nyquist-Johnson relations for *thermal noise*.

Thermal noise can also be expressed as noise power:

$$\langle P_n \rangle = \langle I_n^2 \rangle \cdot R = 4k_B T \Delta f \tag{4.17c}$$

The analog signal is always measured in the presence of noise and the operational limit of analog devices (e.g. the detection limit of sensors) is set by the signal-to-noise ratio (SNR), which is usually defined in terms of a power ratio:

$$SNR = \frac{\langle P_s \rangle}{\langle P_n \rangle} \tag{4.18a}$$

Often it is convenient to express the signal-to-noise ratio through signal amplitude, for example voltage V_s or current I_s. From (4.17a–c) for $R = const.$ obtain:

$$SNR = \left(\frac{V_s}{V_n} \right)^2 = \left(\frac{I_s}{I_n} \right)^2 \tag{4.18b}$$

(V_s, I_s, V_n and I_n are the root mean square values of signal and noise voltage or current amplitudes respectively.)

4.5 WHAT INFORMATION CAN BE OBTAINED FROM CELLS?

The state of a living cell can be monitored by sensing different physical parameters at or near the cell surface [3]. Some examples are given below categorized by the nature of the stimulus.

(i) *Chemical.* Examples of cell chemical signals are ions, such as $\mathbf{H^+}$ or $\mathbf{Ca^{2+}}$, smaller organic molecules, such as sugar, lactones, and larger biomolecules, such as proteins, DNA, etc. One important chemical indicator of the physiological state is pH, which is related to the concentration of hydrogen ions or protons $\mathbf{H^+}$ (see Box 4.2). For example, studies of the mammalian cell's acidification rate (the rate of proton extrusion into extracellular medium) have shown that under steady-state conditions, one cell produces on average ~10^7–10^8 protons ($\mathbf{H^+}$) per second [4]. If a particular function of the cell is activated, the number of extruded protons can increase manyfold. For example, if the antimicrobial function *phagocytosis* is stimulated in a type of white blood cell, *neutrophils*, the acidification rate is increased more than fourfold from 1.3×10^7 H^+/s in the resting state to 5.5×10^7 H^+/s [4].

(ii) *Electrical.* Electrical signaling has a primary importance for the nervous system. Cells that have the ability to generate electrical signals are called *electrogenic* cells (see Box 4.3), such as brain cells (*neurons*) and heart cells. For example, neurons are characterized by 'action potential'

BOX 4.2 PH OF A SOLUTION

pH (from Latin *potentia hydrogeni* – power of hydrogen) is a measure of the *acidity* of a solution. The property of acidity is directly related to the concentration of hydrogen ions $\mathbf{H^+}$ or protons. Quantitatively, it can be approximately expressed as

Table B4.1 Examples of pH of some important liquid constituents of both the exterior and interior of living organisms

Medium	pH
Distilled water	7
Seawater	7.5–8.4
Human blood	7.36–7.44
Human skin	4.5–6.0
Human saliva	5.6–7.6
Gastric juice (acid)	1–2

$$pH \approx -\log\left[H^+\right]$$

where $[H^+]$ is the molar concentration of dissolved hydrogen ions.

In living systems, pH is tightly regulated and is an important measure of the physiological state and, in many cases, an indicator of health. Table B4.1 contains examples of pH of some important liquid constituents of both the exterior and interior of living organisms.

BOX 4.3 ELECTRICITY AND THE CELL

The discovery of the mechanisms underlying cellular electrical activities was led by pioneering work by Hodgkin and Huxley,[a] who were awarded a Nobel Prize in Physiology for Medicine in 1963.

The electrical elements of living cells are comprised of ion pumps and gated ion channels in the cells' membranes. The electrical activity of a cell is due to conditional opening of the channels in the cell's membrane which allows ions to flow to and from the cell's exterior. The ion pump (ion transporter) is a special protein that moves ions across the membrane. Potassium $\mathbf{K^+}$ and sodium $\mathbf{Na^+}$ ions are primary contributors to the membrane current, although some other ions, e.g. chlorine $\mathbf{Cl^-}$ and calcium $\mathbf{Ca^+}$, play a role. The ion current results in a voltage drop across the cell's membrane (~5 nm thick). For example, in the *cardiac myocytes,* which are cells responsible for generating the electrical signals that control the heart rate, the resting heart cell membranes are charged positively on the outside and negatively on the inside due predominantly to K^+ ion gradient which results in a transmembrane potential of about –90 mV.[b]

Because of the importance of electrical measurements/stimulations in heart and brain functions, it is essential to implement accurate electrical measurements for living systems. There are three main categories of experimental measurements of electric signals in the brain:[c]

1. *Local field potentials* – electric potential sensed in the immediate proximity to neurons using microelectrodes of a size comparable to the cell's size.
2. *Electrocorticogram* – electrical measurements using electrodes ~1 mm in size, placed on the cortical surface.
3. *Electroencephalogram* – electrical measurements using electrodes on the centimeter scale placed on the surface of the scalp.

[a] A. L. Hodgkin and A. F. Huxley, "A quantitative description of membrane current and its application to conduction and excitation in nerve", J. Physiol. 117 (1952) 500–544.

[b] W. A. Tacker and L. A. Geddes, "The laws of electrical stimulation of cardiac tissue", Proc. IEEE 84 (1996) 355–365.

[c] C. Bédard, H. Kröger, A. Destexhe, "Model of low-pass filtering of local field potentials in brain tissue", Phys. Rev. E 73 (2006) 051911.

(intracellular) and 'local field potential', the latter of which is the composite extracellular potential field derived from several hundreds of neurons around an electrode tip. The individual action potentials of neurons have a typical duration of 1 millisecond and intracellular amplitude of 100 mV. The extracellular signals (local field potential) have a frequency of 1–100 Hz and potential of ~5 mV. Electrical measurements at the cell membrane provide essential information on the cell's state. A reliable interface between neurons and electrical sensors is required for neuro-electronic devices, including neuro-prostheses and neuro-computers [5].

Cellular electricity can also be a source of energy in living systems. For example, the *electric eel* has several thousands of specialized cells used to generate electricity (*electrocytes*), each cell producing ~0.1 V. The electrocytes in an electric eel are connected in series and can generate voltage up to ~ 600 V. The operating principle of electrocytes has inspired a recent proposal for a new energy source to power medical implants and other small-scale devices [5].

(iii) *Light.* Some cells are naturally bioluminescent, which means they emit light under certain conditions. For example some strains of bacterium *V. cholerae* are luminescent [7]. Another luminescent bacterium, *Xenorhabdus luminescens*, has been found in human wounds [8]. In addition to naturally luminescent cells, today different light-emitting living cells can be genetically engineered and used as 'microbial biosensors' [9], e.g. for detection of toxic materials. Also, the ability to track the light-emitting cells within an organism provides powerful in vivo imaging technology [10]. In current applications, light emitted from such in vivo reporters is detected by an external photosensor through the skin and other tissues; thus, the sensitivity of the method is severely limited by the tissue absorption. Implementations of nanomorphic cells could dramatically expand the capability of the technology for in vivo imaging of light-emitting probes.

(iv) *Thermal.* Living cells are open thermodynamic systems and they continuously exchange matter, energy, and entropy with the surrounding medium [11]. Therefore, bio-calorimetry, which monitors the thermal parameters of biosystems, is an important methodology as it provides direct information about the physiological state of organisms [11, 12].

4.6 SENSORS OF BIOELECTRICITY

The primary technological challenge in the development of electrical biosensors is that the charge carriers in the living cells are ions in *electrolytes*, but in solid-state devices, the charge carriers are electrons. Therefore, integration between *microionics* and *microelectronics* is needed [13].

The electrical elements of living cells are comprised of ion pumps and gated ion channels in the cells' membranes. The electrical activity of a cell is accompanied by the opening of the channels which allows ions to flow from/to the cell's exterior. The time scale of this 'ionic' event is on the order of milliseconds [14,15]. The current flowing through the ion channels causes a voltage change in the electrolyte, which can be sensed by a FET-type device. To enable the 'electronic–ionic' interface, the FET is fabricated without a traditional metal gate electrode; instead, the ion-conductive electrolyte acts as a controlling gate electrode as shown in Figure 4.5. If an ion-sensitive material such as silicon dioxide is used as a gate insulator, the 'gate-less' FET will sense the change of ion concentration. Such a structure is referred to as an electrolyte–oxide–silicon FET or EOS FET [13–19]. It is also called an ion-sensitive field-effect transistor (ISFET).

To sense extracellular electrical activity, the cell must be brought in proximity to the FET channel, as the extracellular electrolyte plays the role of an FET gate. The FET/transistor hybrid is sometimes

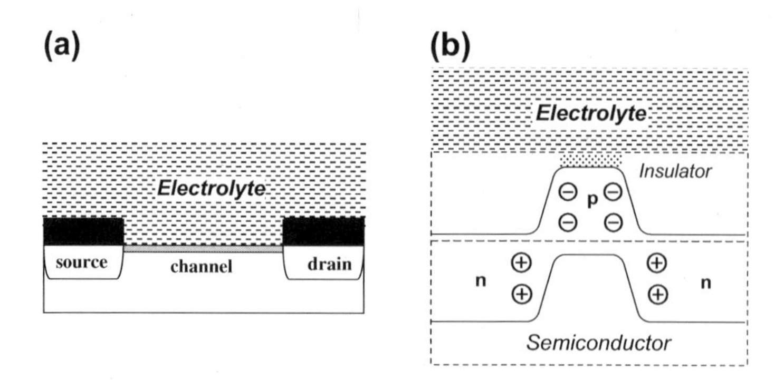

FIGURE 4.5

'Ionic' field-effect transistor: (a) materials system; (b) barrier representation

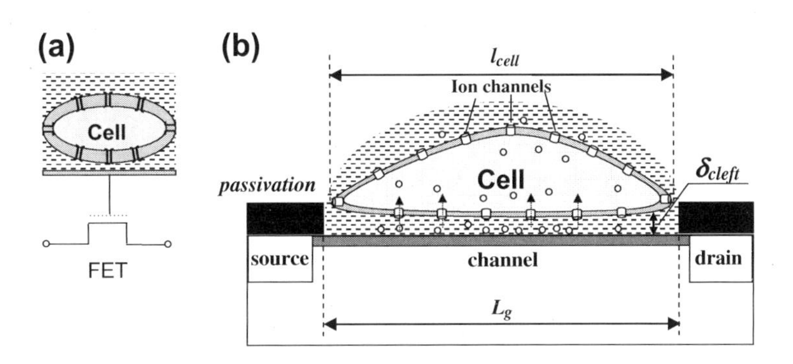

FIGURE 4.6

Recording of electrical activity of an electrogenic cell with FET biosensor: (a) electrical abstraction; (b) schematic of Cell–FET configuration

called a Cell–FET [5]. Such FET biosensors have been successfully used to record electrical signals from individual cells [5,13–20].

A schematic configuration of the Cell–FET is shown in Figure 4.6. Note that there is a gap or cleft in the junction between the cell membrane and FET sensing surface. The cleft is filled with electrolytes. Studies by the Peter Fromherz group at the Max Planck Institute for Biochemistry (Germany) [13,14,16–20] have shown that this thin sheet of electrolytes is an inherent component of interface between a living cell and a solid surface (e.g. of a semiconductor FET sensor). Experiments with rat neurons have shown that a typical thickness of the cleft is δ_{cleft} ~ 50 nm [18].

The scaling sensitivity trade-offs of FET biosensors have been also analyzed in [17, 18]. Based on signal-to-noise ratio (SNR) considerations, it was argued that the optimum 'sensing' area A_g of the FET gate/channel region should be close to the cell size [18]. The arguments are summarized in a simplified form as follows.

Consider a planar Cell–FET with an open gate size L_g and gate surface area $A_g \sim L_g^2$. The FET gate area is in proximity to a cell of size l_{cell} (Fig. 4.6b).

The thermodynamic limit for the signal-to-noise ratio in Cell–FET interfacing is set by the thermal noise [14]. In Section 4.3, the Nyquist-Johnson relation for thermal noise was derived:

$$V_n^2 = 4k_BTR\Delta f \tag{4.15b}$$

Note that the Ohmic resistance of electrolytes in the cleft between the attached cell membrane and the transistor contributes to the thermal noise:

$$R_{cleft} = \rho\frac{d_{cleft}}{A_g} \propto \frac{1}{A_g} \tag{4.19}$$

or

$$V_n^2 \propto \frac{1}{A_g} \sim \frac{1}{L_g^2} \tag{4.20}$$

On the other hand, the Cell–FET gate voltage (signal) is proportional to the gate charge density (charge per gate area):

$$V_s \sim \frac{q}{A_g} \sim \frac{q}{L_g^2} \tag{4.21}$$

The gate charge q is directly proportional to the number of ionic channels, N_{ch}, in the 'sensing' area, i.e. the cell area in contact with the gate. There are two possible cases, depending on the relative sizes of the gate and the cell:

$$l_{cell} < L_g$$

In this case the gate charge is determined by the cell area:

$$q \sim N_{ch} \sim A_{cell} \sim l_{cell}^2 \tag{4.22}$$

Note, the gate charge q remains constant (the entire cell area is exposed) and as a result, the charge density and signal voltage (4.21) decrease as gate area increases:

$$V_s \sim \frac{l_{cell}^2}{L_g^2} \tag{4.23}$$

$$l_{cell} > L_g$$

In this case, the gate charge is directly proportional to the gate area in contact with the cell:

$$q \sim N \sim A_g \sim L_g^2 \tag{4.24}$$

The corresponding signal voltage remains constant, independent of gate size:

$$V_s \sim \frac{q}{A_g} \sim \frac{L_g^2}{L_g^2} = const \tag{4.25}$$

The signal-to-noise ratio of the Cell–FET can be defined as

$$SNR \sim \left(\frac{V_s}{V_n}\right)^2 \tag{4.26}$$

Figure 4.7 illustrates the behavior of input signal and noise in a planar Cell–FET. As can be seen, the signal-to-noise ratio of the Cell–FET reaches a maximum at $L_g = l_{cell}$ ($A_g = A_{cell}$).

Examples of typical gate dimensions of practical FET biosensors are 2 μm × 20 μm, 6 μm × 7 μm, and 22 μm × 24 μm [14, 16]. The RMS noise in the range 2 Hz–2 KHz is ~5 μV for larger transistors and 14 μV for smaller devices [14]. A larger FET allows for 'full-cell' recording, while smaller devices can be used to record, e.g., *vesicle* releases [16]. It is projected that for detection of small vesicles with diameter ~50 nm, a FET sensor with gate area ~0.5 μm^2 will be needed. In this case, the projected signal amplitude is ~400 μV at a detection limit ~300 μV. The latter example could be regarded as an approximate practical scaling limit for planar FET biosensors.

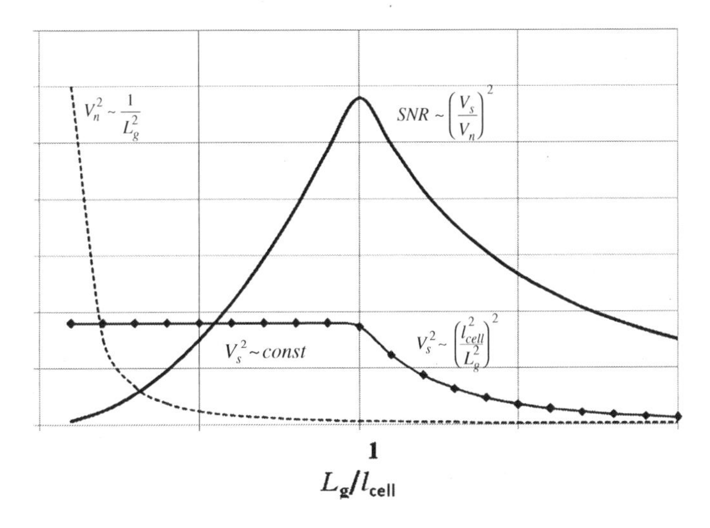

FIGURE 4.7

Scaling behavior of input signal-to-noise ratio of planar Cell–FET biosensor

4.7 CHEMICAL AND BIOCHEMICAL SENSORS

4.7.1 Planar ISFET sensors

An ion-sensitive FET can also be used to analyze the chemical composition of a solution. The simplest goal would be to measure the pH level, which is related to the concentration of hydrogen ions. The ionic activity of the cell can be measured by the ion-sensitive FET (ISFET) described in the previous section. In biological experiments, the ISFETs are often used as a pH sensor measuring concentrations of $\mathbf{H}^+$ ions [21]. Many biochemical reactions are accompanied by changes in pH. For example, bacterial metabolism of sugar in many instances results in a significant pH drop of the medium and can be detected by an ISFET-based biosensor [22].

The ISFET, in its basic configuration, is not selective, i.e. it does not distinguish between different ions or molecules. If ISFETs are used for *selective* detection of, e.g., organic molecules, the FET sensing surface must be covered ('functionalized') by certain chemically selective compounds, for example different ***enzymes*** [21]. If certain target molecules are present in the solution, the enzymes catalyze biochemical reactions that result in a pH change. In a glucose ISFET sensor, the ***glucose oxidase*** enzyme promotes glucose conversion to ***gluconic acid***, which is accompanied by the formation of $\mathbf{H}^+$ ions [23]. Enzyme ISFETs have been developed for the detection of different analytes, for example glucose, penicillin, urea, pesticides, and many others [21].

The ISFET can also be functionalized by other biorecognition molecules – 'molecular receptors', such as single-stranded DNA or proteins [21]. The receptor molecules have high binding affinities to the selected target molecules, as opposed to other molecules in the solution. The target molecules attached to receptors in the ISFET gate area can change barrier height and therefore FET current due to, e.g., molecular dipoles or net charges.

The detection limit of planar FET biosensors is usually in the range of ~10^{-5} mol/L (~10 μM) [21]. Much smaller concentrations are characteristic for individual cell activity, e.g. in the fM to pM range, for nanomorphic configurations (see Box 4.4). Therefore, sensors with lower detection limits are needed.

BOX 4.4 MOLECULAR CONCENTRATIONS IN NANOMORPHIC CONFIGURATIONS

Molar concentration is a measure of the quantity of certain molecules, atoms, or ions in a solution. The molar concentration, μ, is defined as the number of moles per unit volume:

$$\mu = \frac{N_{mol}}{N_A \cdot v}$$

where N_{mol} is the number of molecules of a given type present in the volume, v, of the solution and N_A is Avogadro's Number. A typical unit of molar concentration is 1 molar = 1 M = 1 mol/L.

Consider a nearly spherical cell with a 'radius' r_{cell} ~5 μm. The approximate volume of such a cell is $v_{cell} \sim 4/3\pi r^3$ ~ 5×10^{10} cm^3 = 0.5 pL. Suppose only one biomolecule of a certain type is present in the cell. The corresponding molar concentration is

$$\mu_{cell} = \frac{1}{N_A v_{cell}} \sim 3pM$$

Next, consider a cell that is confined within a closed spherical volume of radius r_s as shown in Figure B4.4a. Suppose the cell expels one molecule to its exterior. The resulting molar concentration for such a closed volume configuration (CVC) is

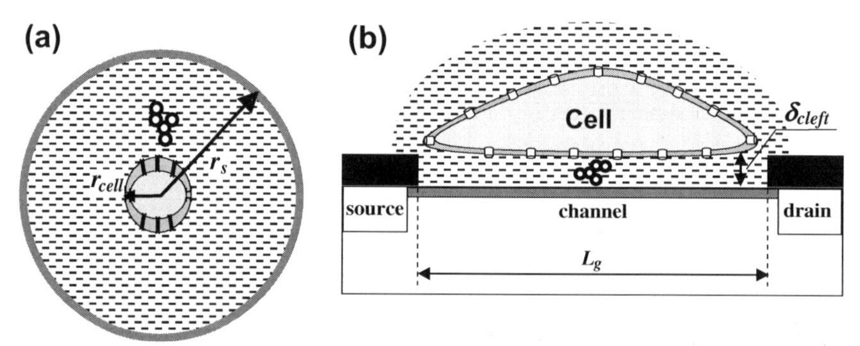

FIGURE B4.4

Nanomorphic configurations for (bio)chemical sensing: (a) closed volume configuration, (b) Cell–FET-like configuration

$$\mu_{cvc} = \frac{1}{N_A} \cdot \frac{3}{4\pi\left(r_s^3 - r_{cell}^3\right)}$$

For example μ_{cvc} = 0.5 pM for r_s = 10 μm and μ_{cvc} = 3 fM for r_s = 50 μm, both for r_{cell} = 5 μm.

Finally, perhaps the smallest possible sampling volume for biochemical sensing is the volume of the 'cleft' in the Cell–FET-like configuration (Fig. B4.4b), which was discussed in Section 4.4. The cleft volume for δ_{cleft} ~ 50 nm and L_g ~ 10 μm is v_{cleft} ~ 5 fL and the corresponding molar concentration is μ_{cleft} ~ 300 pM for one molecule.

4.7.2 One-dimensional nanostructures for biosensing

In 2001, Charles Lieber's group at Harvard demonstrated that using one-dimensional (1D) FET structures (for example Si nanowires) for chemical sensing will dramatically enhance device sensitivity [24]. The sensitivity enhancement by using a nanowire (NW) channel FET is due to the very high surface-to-volume ratio and small cross-sectional area of the nanowire [23–25]. In a conventional (2D) FET, the gate charge changes the channel conductance only in a thin channel-gate insulator interface region, while in NW devices, the conductance is modulated at the 'bulk' level, resulting in higher sensitivity [25]. The ion sensitivity of NW FET increases as the NW radius decreases. Below, a simple analysis to illustrate this point is offered for a p-type doped NW FET pH sensor.

Consider a NW FET structure with a nanowire of diameter $d = 2r$ and length L (Fig. 4.8a). As an example, it is assumed that the NW is made of silicon and is doped by acceptor-type impurities (such as boron) to a concentration N_a to achieve a reasonable conductivity. The current through the nanowire is proportional to the cross-sectional area, A, of the NW channel:

$$I_{NW,0} = \frac{V}{R} = \frac{VA}{\rho L} = \frac{V \cdot \pi r^2}{\rho L} \tag{4.27}$$

(ρ is the resistivity of the nanowire).

When $\mathbf{H}^+$ ions are adsorbed on the NW surface, the NW surface region of width W becomes *depleted* of mobile charge carriers (see Boxes 3.3–5 in Chapter 3). The depleted near-surface region of NW has a low conductivity, and therefore the conductive cross-section A_c of the NW decreases as shown in Figure 4.8b.

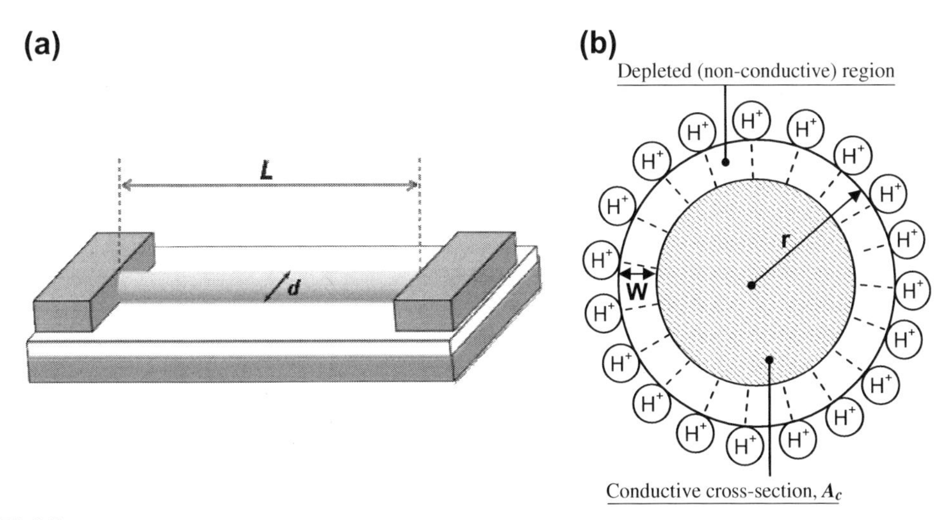

FIGURE 4.8

Nanowire FET sensor: (a) a schematic view; (b) a cross-section of NW showing external surface charge and internal charged depletion layer

The corresponding current through the NW of length L and cross-section area A_c (Fig. 4.8b) is

$$I_{NW} = \frac{V}{R} = \frac{V \bullet A}{\rho L} = \frac{V \bullet \pi (r - W)^2}{\rho L} \tag{4.28}$$

The relative change in current due to the adsorption of ions is a measure of device sensitivity (see Eq. 4.5 in Section 4.3) and this change is given as (assuming $W << r$)

$$\frac{\Delta I_{NW}}{I_{NW,0}} = \frac{r^2 - (r - W)^2}{r^2} = \frac{2rW - W^2}{r^2} \approx \frac{2rW}{r^2} = \frac{2W}{r} \tag{4.29}$$

The depletion width, W, can be estimated based on the charge neutrality condition, i.e. the total positive charge of ions absorbed at the surface, q_s, is equal to the total negative charge, q_d, of ionized acceptors in the depleted volume of the nanowire:

$$\begin{aligned} q_s &= q_d \\ en_s \bullet 2\pi rL &= eN_a \bullet 2\pi rLW \end{aligned} \tag{4.30}$$

where n_s is the surface concentration of absorbed charge (cm^{-2}), and N_a is the volume concentration of ionized acceptor impurities (cm^{-3}).

From (4.28), the depletion width W can be found as

$$W = \frac{n_s}{N_a} \tag{4.31}$$

Putting (4.31) into (4.29), the NW FET sensor sensitivity is

$$\frac{\Delta I_{NW}}{I_{NW,0}} \approx \frac{2}{r}\frac{n_s}{N_a} \tag{4.32}$$

Because of its high sensitivity, the nanowire FET shows promise for both chemical and biochemical microsensors. Si-NW FETs for pH sensing have been demonstrated in [23–26]. Typical NW sensor dimensions are ~50–100 nm in diameter and several μm in length [23–26]. The conductance of the silicon nanowires has been shown to be directly related to the pH of the surrounding environment. For example, more than one order of magnitude in current change resulted by varying pH values from 6.0–8.0 [26]. Smaller devices exhibited even greater sensitivity. In [25], the NW pH sensor was tested in the pH range 2–12 and a sensitivity of ~80 nS/pH was achieved.

The Si-NW FET pH sensors have been used, for example, to monitor real-time cellular responses – specifically, *T-lymphocyte* activation. Addition of a species-specific *antibody* to a suspension containing *T-cells* results in the release of acid by the cells, which causes current decreases in the NW sensor [26].

While nanostructured silicon is a popular material for biochemical sensing applications, some other materials also show promise. Carbon nanotubes (CNT) have attracted interest in biosensing applications [27, 28]. For example, CNT biosensors were used for real-time detection of *poly-L-lysine* with a detection limit of ~1 pM [27].

The CNT diameter can be as small as 1 nm (comparable with the size of a single protein), while a typical length of the CNT for biosensing applications is 1–100 μm, and thus the CNT can span a single cell. In principle, they could be used as (quasi) non-invasive probes to contact or even puncture

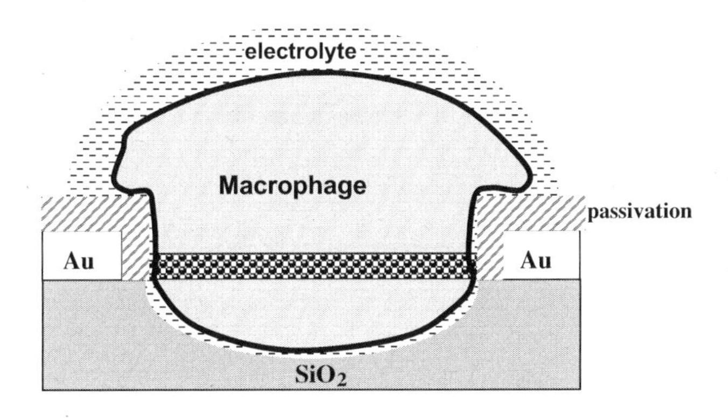

FIGURE 4.9

Suspended CNT sensor interfacing with a macrophage cell

(adapted from [28])

the cell's membrane [28]. In an interesting recent experiment by the Cees Dekker's group at Delft University, a suspended CNT FET sensor was used to monitor a *macrophage.* (Macrophages play an important role in the human immune system by engulfing and ingesting *pathogens.*) In [28], a FET sensor, equipped with a suspended nanowire channel, was utilized to interrogate single macrophage cells. The structure of the suspended 1D channel FET allowed the macrophage to surround the channel (Fig. 4.9) as it is naturally 'programmed' to do. This preliminary result suggests an intriguing possibility of a macrophage cell ingesting an electrical sensor and thus the possibility of electrically monitoring processes inside the cell [28].

Another application for one-dimensional FET sensors is the direct detection of biological macromolecules. As was discussed above, this application requires that the nanowire be functionalized by binding, e.g., a receptor molecule to the surface of a NW sensor. NW FETs functionalized with antibodies were successfully utilized for detection of antigens (immunodetection) at less than 100 fM concentrations [26]. As another example, NW FET sensors were functionalized with *biotin* for selective detection of *streptavidin* proteins. Concentrations of the *streptavidin* molecules as low as 10 fM were detected (at SNR of 140) and the detection floor was projected to be ~70 aM [26].

The detection limit of nanobiosensors has also been the subject of recent theoretical studies by the group of Alan Alam at Purdue University, who used Monte Carlo simulations to connect detection limits with the time required for detection. It was reported that experimentally measured concentrations may be influenced by statistical effects [29,30].

4.8 THERMAL BIOSENSORS

4.8.1 Basic principles

Monitoring the thermal environment in very small volumes, e.g. around the living cells, can also provide important information about their physiological state; however, the small size of the cells places a number of constraints on the design and operation of thermal sensors. For example, as was discussed in

Chapter 1, typical bacteria whose dimensions are approximately 1 μm, have a volume ~10^{-12} cm^3 (or ~1 fL) and metabolic power P_{cell} ~ 10^{-16}–10^{-13} W. On the other hand, a typical human cell is of the size ~10 μm, with a volume 10^{-9} cm^3 (or 1 pL) and metabolic power P_{cell} ~ 10^{-12}–10^{-10} W.

Below, some scaling considerations for bio-thermal sensors are addressed. First, a well-known relation in thermal physics relates the amount of heat, Q, released or absorbed within a physical system to the corresponding temperature change ΔT:

$$Q = C_{th}\Delta T = c_{vol} \cdot v \cdot \Delta T \tag{4.33}$$

where C_{th} is the 'thermal mass' or thermal capacitance, c_{vol} is the *volumetric* heat capacity, and v is the thermometric volume. Table 4.1 shows values of c_{vol} for different substances.

Note that the volumetric heat capacity for most solid and liquid substances lies in the interval 1.5–4 J/$cm^3 \cdot$K (water's heat capacity of 4.2 J/$cm^3 \cdot$K is the largest of all known substances). Therefore in the analysis below, an 'average' volumetric heat capacitance is assumed to be:

$$c_{vol} \sim const \sim 3 \frac{J}{cm^3 \cdot K}$$

Next, from (4.33), the temperature change due to the release/absorption of heat Q is

$$\Delta T = \frac{Q}{c_{vol} \cdot v} \approx \frac{Q}{3v} \tag{4.34a}$$

Note that in thermal measurements, sensor components add to the thermometric volume and it may also generate additional heat as a result of its operation; thus

$$Q = Q_a + Q_s$$

$$v = v_a + v_s$$

where Q_a and v_a are respectively the heat and volume of the sensor 'addenda', and Q_s and v_s are the sampling heat and volume.

$$\Delta T = \frac{1}{3}\frac{Q_a + Q_s}{v_a + v_s} \tag{4.34b}$$

Table 4.1 Volumetric heat capacity for different substances

Substance	$c_{vol}, \frac{J}{cm^3 \cdot K}$
Water	4.2
Human brain tissue	~3.7
Copper	3.5
Gold	2.5
Zinc	2.8
Silicon	1.7
Silicon oxide (fused silica)	1.5
Silicon nitride	2.4
Polyethylene	~2.1

Clearly, for accurate thermal measurements, the influence of the sensor addenda must be minimized, i.e. $Q_a << Q_s$ and $v_a << v_s$. Therefore zero-power sensing ($Q_a \sim 0$) is desirable at a small scale, where the input heat energy directly generates an output electrical signal. Also, for nano-bio-thermometry it is critical that the sensor thermal mass is small compared to the measured sample, i.e. the sensor volume should be minimized.

4.8.2 FET-type thermal sensors

In principle, a FET-like barrier structure can be used for temperature measurements. As was extensively discussed in this and previous chapters, the current through a barrier structure (e.g. of Fig. 4.4) is

$$I = eN_0 \exp\left(-\frac{E_b}{k_B T}\right)\left(1 - \exp\left(-\frac{V_{AB}}{k_B T}\right)\right) \tag{4.35}$$

Since the current (4.35) has a temperature dependence, it is possible to determine temperature from current measurements. Suppose the sensor current is I_0 at $T = T_0 = 300$ K and $I = I_1$ at certain temperature $T_1 = T_0 + \Delta T$, where ΔT is a small temperature increment. From (4.35) we obtain:

$$\frac{I_1}{I_0} \approx \exp\left(\frac{E_b}{k_B}\left(\frac{1}{T_0} - \frac{1}{T_1}\right)\right) \tag{4.36}$$

(In (4.36) it is assumed that the number of electrons that strike the barrier per unit time, N_0, is temperature independent. In reality, N_0 depends on the temperature; however, it is easy to show that for small $\Delta T = T_1 - T_0$ and sufficiently high barriers, one can assume $N_0 \approx$ const.)

A simple transformation of (4.36) yields:

$$\frac{k_B}{E_b}\ln\left(\frac{I_1}{I_0}\right) \approx \frac{1}{T_0} - \frac{1}{T_1} \tag{4.37a}$$

Now requiring $I_1/I_0 = 2$ (as was discussed in Section 4.3, see Eq. 4.5) and $T_0 = 300$ K we obtain:

$$T_{1_{\min}} \approx \left(\frac{1}{300} - \frac{k_B}{E_b}\ln 2\right)^{-1} \tag{4.37b}$$

For $E_b = 1$ eV, $T_{1min} = 305.48$ K, and therefore $\Delta T = 5.48$ K.

Figure 4.10 shows values of ΔT for E_b values in the range 0.5–1 eV (typical for silicon devices), calculated using formula (4.37). For comparison, more accurate calculations taking into account temperature dependence of N_0 in (4.35) are also shown (dashed line).

From the above discussion, the measurement of temperature-dependent current through a barrier structure (or in other words the *resistance* of the structure) provides a useful method for thermal measurements. Indeed, a broad class of temperature sensors is based on the resistance-based devices, such as thermistors, semiconductor FET- and diode-based thermal sensors. However, these devices in general have limited sensitivity [31] and they may not be sufficiently sensitive for small-scale bio-measurement, where typical temperature variations are less than 1 K. Another drawback of the resistance-based thermal sensors is their heat generation during operation as a result of current flow. As was discussed in Section 4.6.1, this is an undesirable effect in small-scale temperature measurements. Nevertheless there are attempts to utilize semiconductor barrier

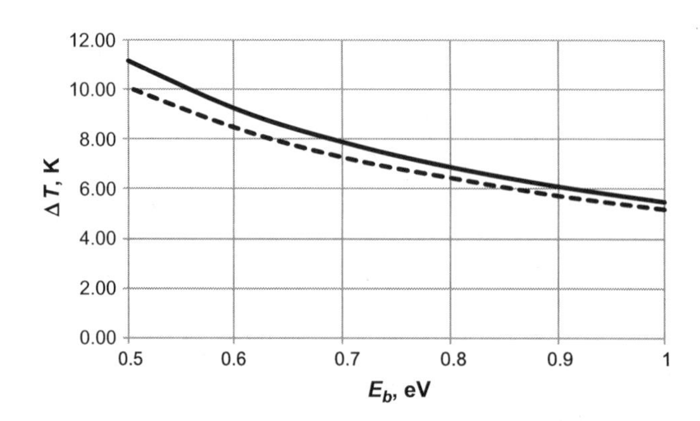

FIGURE 4.10

Temperature sensitivity of a barrier structure: minimum temperature increment ΔT required for a $I_1/I_O = 2$. Solid line represents approximate model (Eq. 4.37) and the dashed line, more accurate calculations taking into account temperature dependence of N_O in (4.35)

structures, such as Schottky barriers (see Box 3.4 in Chapter 3), for small-scale thermal measurements [32].

4.8.3 Thermoelectric sensors

Zero-power sensing ($Q_a \sim 0$) is desirable at a small scale, where the input heat energy directly generates an output electrical signal. Such thermoelectric transducers (thermocouples) are described in Box 4.5 and they are the typically used for thermal measurements from very small objects [33–37].

To estimate the detection limit of cell thermometry, consider the configuration of Figure 4.11. The sampled cell is located near the surface of the nanomorphic cell, separated by a 'cleft'. Such a configuration is similar to the Cell–FET-like configuration, discussed in Sections 4.4 and 4.5. (The cleft volume for $\delta_{cleft} \sim 50$ nm and $L \sim 10$ µm is $v_{cleft} \sim 5 \times 10^{-12}$ cm^3 = 50 fL.)

With a metabolic power of cell P_{cell}, the amount of heat released in the cleft volume v_{cleft} during sampling time t_s is

$$Q_s \sim \frac{1}{2} P_{cell} \cdot t_s \tag{4.38}$$

(the ½ factor appears because in the nanomorphic configuration of Figure 4.11, only a half of the cell is in contact with the measuring device).

The temperature change in the cleft due to heat Q_s is given by (4.34) and the signal voltage generated by the thermocouple is

$$V_s = S_T \Delta T = \frac{S_T}{3} \frac{Q_s}{v_a + v_s} \tag{4.39}$$

where S_T is the Seebeck coefficient of the thermocouple (see Box 4.5).

BOX 4.5 THERMOELECTRICITY

As was discussed in Chapter 3 (see Boxes 3.4 and 3.5), when two different media are brought in contact, a barrier is formed at the interface due to different charge concentrations. For example, in a junction between two different metals, a barrier, known as a contact potential difference (CPD), is formed as a result of different electron concentrations. Such metal junctions are used as thermoelectric sensors (and more generally as energy converters), known as thermocouples. A thermocouple is made from two dissimilar metals **Me₁** and **Me₂** joined together, as shown in Figure B4.5, to form two junctions: junction **1** and junction **2**. The two metals have different concentrations of electrons, respectively n_{e1} and n_{e2}, and there are CPD barriers, E_b, at both junctions. Next, at a given temperature (e.g. T_1 at junction **1** and T_2 at junction **2**), there will be an over-barrier electron flow in both directions. The electron flow depends on the number of electrons that strike the barrier per unit time. Note that the number of electrons is now different at each side of the barrier (dissimilar metals); for a symmetrical barrier (barrier height is the same at both sides), there would be a preferential current supply from the side with the higher concentration of electrons. Since at equilibrium the net current must be zero, the barrier shape is deformed to create the asymmetry needed to accommodate the difference in electron supply. As a result, the barrier at the side with the higher supply will be higher by a certain amount, $e\varphi_1$, as follows from the zero net current condition:

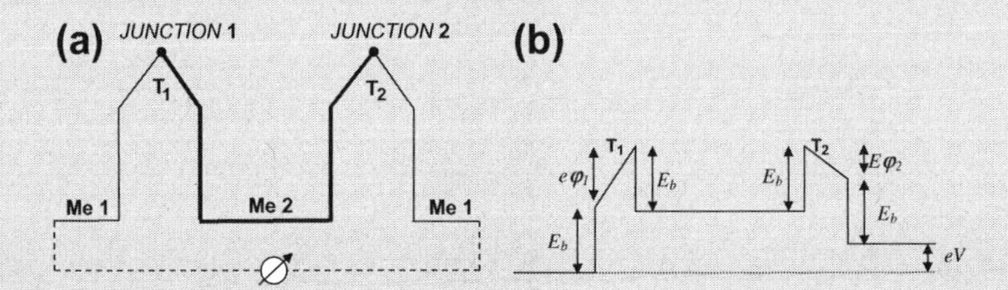

FIGURE B4.5

Thermocouple: (a) physical structure; (b) corresponding barrier model

$$N_1 \exp\left(-\frac{E_b}{k_B T_1}\right) = N_2 \exp\left(-\frac{E_b + e\varphi_1}{k_B T_1}\right) \tag{B4.1}$$

Note that N_1 and N_2 represent the number of electrons available at the barrier interface, i.e. two-dimensional electron concentration, which, in turn, is related to the 3D bulk electron concentration in metal n_e as follows:

$$N_{2D} \sim n_e \tag{B4.2}$$

From (B4.1) and (B4.2), the potential shift at junction **1** (at temperature T_1) is:

$$\varphi_1 = \frac{2}{3}\frac{k_B}{e} T_1 \ln\frac{n_{e2}}{n_{e1}} \tag{B4.3a}$$

Similarly for junction **2** (at temperature T_2):

$$\varphi_2 = \frac{2}{3}\frac{k_B}{e} T_2 \ln\frac{n_{e2}}{n_{e1}} \tag{B4.3b}$$

And the overall potential difference (voltage) across the ends of the thermocouple circuit is:

$$V = \varphi_2 - \varphi_1 = \left(\frac{2}{3}\frac{k_B}{e}\ln\frac{n_{e2}}{n_{e1}}\right)\Delta T = S_T \Delta T \tag{B4.4}$$

where S_T is 'thermopower' or the Seebeck coefficient of the thermocouple. It is a measure of the sensitivity of the thermocouple temperature sensor in V/K.

As a numerical example, consider the Ni–Au thermocouple discussed in this chapter. Electron concentration in metals is approximately equal to their atomic concentration, $n_e \sim n_{at}$. Thus $n_{eAu} \sim 6 \times 10^{22}$ cm^{-3}, $n_{eNi} \sim 9 \times 10^{22}$ cm^{-3}, and from (B4.4) we obtain:

$$V_{Ni-Au} \sim \left(23\,\frac{\mu V}{K}\right) \cdot \Delta T$$

which is close to the experimental value of $S_T = 22.5$ μV/K for thermoelectric power of the Ni–Au thermocouples[a]. In order to increase the voltage output, several thermocouples are often connected in series, forming a *thermopile* (TP).

[a] D. D. Thornburg and C. M. Wayman, "Thermoelectric power of vacuum-evaporated Au-Ni thin-film thermocouples", J. Appl. Phys. 40 (1969) 3007–3013

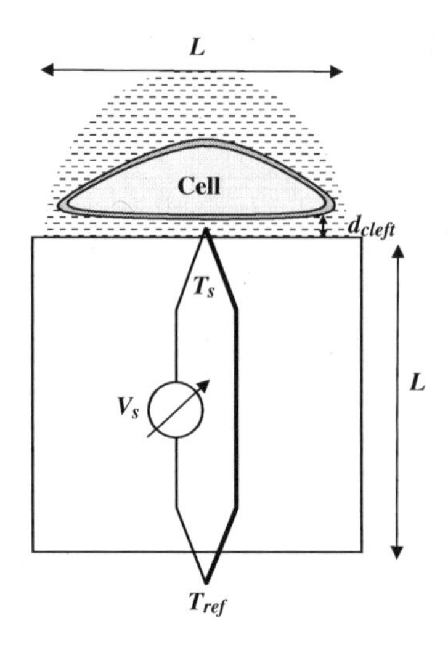

FIGURE 4.11

'Minimal configuration' for cell thermometry: The sampled cell is located near the surface of the nanomorphic cell

The sampling volume v_s is the volume formed by the cleft:

$$v_s = L^2 \cdot \delta_{cleft} \tag{4.40a}$$

Suppose the thermocouple is formed by two metal wires of length L and cross-sectional dimension w. Then the volume of the addenda is assumed to be half of the total volume of the thermocouple:

$$v_a = Lw^2. \tag{4.40b}$$

Thus, the signal voltage from (4.39) is

$$V_s = \frac{S_T}{3} \frac{P_{cell} t_s}{(Lw^2 + L^2 \delta_{cleft})} \tag{4.41}$$

As was discussed in previous sections, the thermodynamic limit for the signal detection set by the thermal noise:

$$V_n^2 = 4k_B TR\Delta f \tag{4.15b}$$

where $\Delta f \sim 1/t_s$ and R is the resistance of the thermocouple wires: $R = \rho \frac{2L}{w^2}$ and ρ is the mean resistivity of the thermocouple wires (see Table 4.2).

The signal-to-noise ratio of the sensor is

$$SNR = \left(\frac{V_s}{V_n}\right)^2 = \frac{(S_T P_{cell} t_s)^2}{9(Lw^2 + L^2 \delta_{cleft})^2} \frac{1}{4k_B TR\delta f} = \frac{S_T^2}{72k_B T} \frac{P_{cell}^2 t_s^3}{\rho} \frac{w^2}{L(Lw^2 + L^2 \delta_{cleft})^2} \tag{4.42}$$

Table 4.2 Parameters for scaling analysis of bio-thermal measurements

Parameter	Numerical value	Rational
Metabolic power of the cell, P_{cell}	10^{-12} W (1 pW)	For different single-cell organisms P_c can be in the range 10^{-16}–10^{-10} W
Ambient temperature, T	300 K	Numerical convenience for 'room temperature'
Volumetric heat capacity of sampling substances, c_{vol}	3 J/cm^3·K	For most solid and liquid substances c_{vol} lies in the interval 1.5– 4 J/cm^3·K – see Table 4.1
Cleft between the cell and the base of the sensor, δ_{cleft}	50 nm	An inherent gap at the interface between living cell and a solid surface [18]
Spatial dimension of the base of the sensor, L	10 μm	Nanomorphic cell size
(Min.) length of thermocouple wires, L	10 μm	Nanomorphic cell size
(Max.) Width of thermocouple wires, w	<10 μm	Nanomorphic cell size
Resistivity of thermocouple wires, ρ	5 μΩ·cm	Approx. average for **Ni–Au** system: ρ_{Ni} = 6.9 μΩ·cm, ρ_{Au} = 2.2 μΩ·cm
Thermocouple's sensitivity, S_T	20 μV/K	Approx. value for **Ni–Au** TC (see Box 4.5)
Sampling volume	5 x 10^{-12} cm^3 (5 fL)	$v_s = L^2 \cdot d_{cleft}$

Note that w and t_s are the only two variables in (4.42).

The signal-to-noise ratio of the thermocouple sensor in the nanomorphic configuration as a function of the cross-sectional dimensions of the thermocouple wires is shown in Figure 4.12 for different values of t_s. As can be seen from Figure 4.12, the optimal cross-sectional dimension w of the thermocouple wires is ~0.7 μm, and thus the total volume of the thermocouple is $2Lw \approx 10\ \mu m^3 = 10^{-12}\ cm^3$.

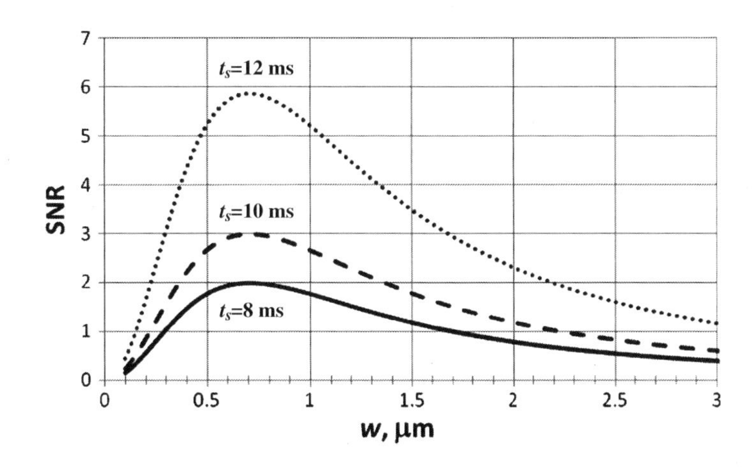

FIGURE 4.12

Signal-to-noise ratio of the thermocouple sensor in the nanomorphic configuration as a function of the cross-sectional dimensions of the thermocouple wires

Table 4.3 State of the art of chip-based calorimeters for nanothermal measurements (adapted from [33])

Group	Volume	Thermal sensor	Power resolution	Comment
CalTech [33]	3.5 nL	**Ni–Au** TP*	4.2 nW	Monitoring of urea hydrolysis and water/methanol mixing
TU Freiberg [12]	6 μL	**BiSb–Sb** TP*	10 nW	Monitoring of metabolic activities of bacterial biofilms, bacteria cultured in suspensions, and single fish embryos
U. Glasgow [35]	720 pL	**Ni–Au** TP*	13 nW	Monitoring of metabolic activity of a single living cell
Vanderbilt U. [36]	<500 pL	**Ti–Bi** TP*	150 nW	Monitoring of evaporation of water droplets
PennState [31]	15 nL	**Si–Au** TP*	300 nW	Monitoring of enzymatic reactions: glucose oxidation and urea hydrolysis

TP = thermopile.

4.8.4 Remarks on the state of the art of nano-biocalorimetry

Table 4.3 summarizes examples for small-volume thermal monitoring of biological and chemical processes. Thin-film thermocouples are commonly used as sensors in these applications. Remarkable progress in on-chip nano-biocalorimetry has been achieved in recent years, and further developments might make nanomorphic thermal measurements plausible. According to the analysis presented in this section, nano-thermal measurements, in principle, could be performed with sufficient sensitivity and time resolution to detect the state of a living cell. Moreover, it appears that for a number of tasks, thermal measurements might be a preferable way to obtain critical information. One example is detection of alive and dead cells at different scales. In fact, recently, a group at TU Freiberg introduced chip calorimetry as a new monitoring tool to gather information about the physiological state of biofilms [12,37]. Formation of biofilms, which are aggregates of microorganisms, such as bacteria, may cause serious medical problems, as biofilms are involved in many infectious diseases in the body. Biofilm formation on implanted prosthetic devices is also an important issue. Biofilm cells show increased resistance to antimicrobial agents, such as antibiotics, and therefore fast and quantitative monitoring of the effects of antimicrobial treatments on the physiological state of biofilms is important. Chip calorimetry has demonstrated the potential to perform economical and fast investigations of the effects of biofilm treatment [37].

4.9 CONCLUDING REMARKS

In this chapter, several types of sensors were discussed: (bio)electrical, (bio)chemical, and (bio) thermal. In spite of the very different natures of stimuli, all these sensors were discussed within the framework of the barrier model, which is a useful abstraction to analyze scaling limits of sensors. Similar analyses based on the barrier model can be performed for other types of sensors, for example light or pressure sensors; unfortunately, the size and scope of this book does not allow for a complete study of sensors.

A 'nanomorphic configuration' was assumed in the analyses of sensor operation throughout this chapter, which implies lateral dimensions of the sensing surface of ~10 μm and an inherent gap ('cleft') of ~ 50 nm at the interface between the living cell and the sensing surface. The volume of such a 'cleft' is very small, ~5 fL, and such a small volume might make the task of, e.g., biochemical and bio-thermal measurements easier compared to test-tube scale measurements. For example, the lower concentration range for biochemical sensing would be ~100 pM, which is within the experimentally reported detection capabilities for biochemical sensors, utilizing one-dimensional structures.

It appears that the 1D structures, such as semiconductor nanowires and carbon nanotubes, might be essential for nanomorphic biosensing, as they have better sensitivity than planar devices and allow for the picomolar detection of biomolecules. In addition, 1D structures with very small diameters, such as carbon nanotubes, could be used as (quasi) non-invasive probes to contact or even puncture the cell's membrane or be 'ingested' by a cell. This suggests an intriguing possibility of electrically monitoring processes inside the cell.

The discussion offered in this chapter considers only a few of the basic steps involved in bio-detection, mainly related to the signal transduction. There are other steps, which were not covered here, including: (i) pre-sensing steps: extraction, separation, amplification and (ii) post-sensing steps: read-out, signal conditioning, and data processing. The latter may need special attention as the power consumption due to sensing operation is dominated by downstream analog circuits.

In general, sensing of different physical stimuli (e.g. electrical, chemical, or thermal) appears to be plausible in nanomorphic configurations, with sufficient sensitivity and time resolution to detect the state of a living cell. It appears that for a number of tasks, thermal measurements might be a preferable way to obtain critical information. One example is detection of alive and dead cells at different scales.

Understanding mechanisms, by which living cells interact with the external world, not only allows one to monitor cell activities, but also could inspire broader developments. For example, the mechanisms underlying cellular electrical activities, such as ion pumps and gated ion channels, might also provide operating principles for a new efficient energy source to power medical implants and other small-scale devices [6].

GLOSSARY OF BIOLOGICAL TERMS

Antibody: A specialized immune protein produced because of the introduction of an *antigen* into the body and which possesses the remarkable ability to combine with the very antigen that triggered its production. Antibodies are also called *immunoglobulins*.

Antigen: *Anti*body *gen*erator – a molecule that triggers the production of antibodies that will bind to the antigen.

Biofilm: An aggregate of microorganisms in which cells are stuck to each other.

Biotin: Vitamin B_7

Electrogenic cells: Cells that have the ability to generate electrical signals. Examples are *neurons* (brain cells), *cardiac myocytes* (heart cells), and specialized 'electrical power' cells (*electrocytes*) used in electrical organs of electric fish.

Enzymes: Proteins that promote (catalyze) (bio)chemical reactions. For example, the *glucose oxidase* enzyme promotes glucose ($C_6H_{12}O_6$) conversion to *gluconic acid* ($C_6H_{12}O_7$).

In vivo: Refers to a biological study that takes place within the living organism.

Lymphocytes: A type of small white blood cell that plays a large role in defending the body against disease. There are two types of lymphocytes:
B cells make antibodies to attack bacteria and toxins.
T cells attack body cells that have been taken over by viruses or have become cancerous.

Macrophages: A type of large white blood cell. Macrophages play an important role in the human immune system by engulfing and ingesting infectious agents (*pathogens*). Human macrophages are about 20 μm in size.

Polylysine: (poly-L-lysine) is a small polymer of the amino acid L-lysine that is produced by bacterial fermentation.

Streptavidin: A type of protein widely used in molecular biology for detection of various biomolecules.

Vesicle: A relatively small intracellular, membrane-enclosed reservoir that stores substances.

LIST OF SYMBOLS

Symbol	Meaning
A	Area
c_{vol}	Volumetric heat capacity
C_{th}	Thermal capacitance
d	Diameter
e	Electron charge, $e = 1.6 \times 10^{-19}$ C
E_b	Energy barrier height
f	Frequency
I, i	Current
k	Number of trials
k_B	Boltzmann constant, $k_B = 1.38 \times 10^{-23}$ J/K
L, l	Length
l_{cell}	Cell size
n_{at}	Atomic concentration
n_e	Concentration of electrons
n_s	Surface charge concentration
N_0	Number of electrons that strike barrier per unit time
N_a	Concentration of ionized acceptor impurities in semiconductor NW
N_A	Avogadro's Number, $N_A = 6.022 \times 10^{23}$ mol^{-1}
N_{mol}	Number of molecules

Symbol	Meaning
P	Power
P_{cell}	Cell's metabolic power
q	Electric charge
Q	Heat energy
r	Radius
R	Resistance
SNR	Signal-to-noise ratio
$s(t)$,s, $\bar{s}_\tau$	Analog signal quantity
S_T	Seebeck coefficient
t	Time
T	Absolute temperature
v	Volume
V	Voltage
w	Cross-sectional dimension/width
W	Depletion width
δ_{cleft}	Cleft thickness
Δf	Bandwidth
μ	Molar concentration
Π	Probabilities of 'success' in one trial
$\pi(k)$, π	Probabilities of 'success' in k trials
ρ	Resistivity
τ	Time interval, sampling time
φ	Electric potential
$\propto$	Indicates proportionality
$\sim$	Indicates order of magnitude

References

[1] D. Grieshaber, R. MacKenzie, J. Vörös, E. Reimhult, Electrochemical Biosensors – Sensor Principles and Architectures, Sensors 8 (2008) 1400–1458.

[2] L. Callegaro, Unified derivation of Johnson and shot noise expressions, Am. J. Phys. 74 (2006) 438–440.

[3] L. Bousse, Whole cell biosensors, Sensors and Actuators B 34 (1996) 270–275.

[4] J.C. Owicki, J.W. Parce, Biosensors based on the energy metabolism of living cells: The physical chemistry and cell biology of extracellular acidification, Biosensors & Bioelectronics 7 (1992) 255–272.

[5] A. Poghossian, S. Ingebrandt, A. Offenhausser, M.J. Schöning, Field-effect devices for detecting cellular signals, Seminars in Cell & Developmental Biology 20 (2009) 41–48.

[6] J. Xu, D.A. Lavan, Designing artificial cells to harness the biological ion concentration gradient, Nature Nanotechnology 3 (2009) 666–670.

[7] C.J. Grim, E. Taviani, M. Alam, A. Huq, R.B. Sack, R.R. Colwell, Occurrence and expression of luminescence in *Vibrio cholerae*, Appl. and Environ. Microbiol. 74 (2008) 708–715.

[8] P. Colepicolo, K.-W. Cho, G.O. Poinar, J.W. Hastings, Growth and luminescence if the bacterium *Xenorhabdus luminescens* from a human wound, Appl. and Environ. Microbiol. 55 (1989) 2601–2606.
[9] S.F. D'Souza, Microbial biosensors, Biosensors & Bioelectronics 16 (2001) 337–357.
[10] B.W. Rice, M.D. Cable, M.B. Nelson, *In vivo* imaging of light-emitting probes, J. Biomed. Optics 6 (2001) 432–440.
[11] J.-L. Garden, H. Guillou, A.F. Lopeandia, J. Richard, J.-S. Heron, G.M. Souche, et al., Thermodynamics of small systems by nanocalorimetry: From physical to biological nano-objects, Thermochimica Acta 492 (2009) 16–28.
[12] J. Lerchner, A. Wolf, H.-J. Schneider, F. Mertens, E. Kessler, V. Baier, et al., Nano-calorimetry of small-sized biological samples, Termochimica Acta 477 (2008) 48–53.
[13] P. Fromherz, Three Levels of Neuroelectronic Interfacing: Silicon chips with ion channels, nerve cells, and brain tissue, Ann. N.Y. Acad. Sci. 1093 (2006) 143–160.
[14] M. Voelker, P. Fromherz, Nyquist noise of cell adhesion detected in a neuron-silicon transistor, Phys. Rev. Lett. 96 (2006) 2281102.
[15] H. Lee, D. Ham, R.M. Westervelt (Eds.), CMOS Biotechnology, Springer, 2007.
[16] J. Lichtenberger, P. Fromherz, A cell-semiconductor synapse: Transistor recording of vesicle release in chromaffin cells, Biophysical J. 92 (2007) 2262–2268.
[17] M. Voelker, P. Fromherz, Signal transmission from individual mammalian nerve cell to field-effect transistor, Small 1 (2005) 206–210.
[18] M. Voelker, Neurochips: field-effect devices for communication with cells, SRC/NSF Forum on Nano-Morphic Systems: Processes, Devices, and Architectures, CA, Stanford University, Stanford, November 8–9, 2007.
[19] P. Fromherz, Joining microelectronics and microionics: Nerve cells and brain tissue, Solid-State Electron 52 (2008) 1364–1373.
[20] I. Peitz, P. Fromherz, Electrical interfacing of neurotransmitter receptor and field effect transistor, Eur. Phys. J. E 30 (2009) 223–231.
[21] C.-S. Lee, S.K. Kim, M. Kim, Ion-sensitive field-effect transistor for biological sensing, Sensors 9 (2009) 7111–7131.
[22] M. Castellarnau, N. Zine, J. Bausells, C. Madrid, A. Juarez, J. Samitier, et al., ISFET-based biosensor to monitor sugar metabolism in bacteria, Mat. Sci. Eng. C28 (2008) 680–685.
[23] E. Stern, A. Vacic, M.A. Reed, Semiconducting nanowire field-effect transistor biomolecular sensor, IEEE Trans. Electron. Dev. 55 (2008) 3119–3130.
[24] Y. Cui, Q. Wei, H. Park, C.M. Lieber, Nanowire nanosenosrs for highly sensitive and selective detection of biological and chemical species, Science 293 (2001) 1289–1292.
[25] I. Park, Z.Y. Li, X. Li, A.P. Pisano, R.S. Williams, Towards the silicon nanowire-based sensor for intracellular biochemical detection, Biosensors and Bioelectron. 22 (2007) 2065–2070.
[26] Eric Stern, J.F. Klemic, D.A. Routenberg, P.N. Wyrembak, D.B. Turner-Evans, A.D. Hamilton, et al., Label-free Immunodetection with CMOS-Compatible Nanowires, Nature 445 (2007) 519–522.
[27] J.N. Tey, I.P.M. Wijaya, Z. Wang, W.H. Goh, A. Palaniappan, S.G. Mhaisalkar, et al., Laminated, microfluidic-integrated carbon-nanotube based biosensors, Appl. Phys. Lett. 94 (2009) 0131107.
[28] I. Heller, W.T. T Smaal, S.G. Lemay, C. Dekker, Probing macrophage activity with carbon-nanotube sensors, Small 5 (2009) 2528–2532.
[29] P.R. Nair, M.A. Alam, Performance limits of nanobiosensors, Appl. Phys. Lett. 88 (2006) 233120.
[30] J. Go, M.A. Alam, Statistical interpretation of 'femptomolar' detection, Appl. Phys. Lett. 95 (2009) 033110.
[31] L. Wang, D.M. Sipe, Y. Xu, Q. Lin, A MEMS thermal biosensor for metabolic monitoring applications, J. Microelectromech. Syst. 17 (2008) 318–327.

[32] T.K. Hakala, J.J. Toppari, P. Törmä, A hybrid method for calorimetry with subnanoliter samples using Schottky junctions, J. Appl. Phys. 101 (2007) 034512.
[33] W. Lee, W. Fon, B.W. Axelrod, M.L. Roukes, High-sensitivity microfluidic calorimeters for biological and chemical applications, PNAS 106 (2009) 1525–15230.
[34] Y. Zhang, S. Tadigadapa, Calorimetric biosensors with integrated microfluidic channels, Biosensors and Bioelectronics 19 (2004) 1733–1743.
[35] E.A. Johannessen, J.M.R. Weaver, P.H. Cobbold, J.M. Cooper, Heat conduction nanocalorimeter for pL-scale single-cell measurements, Appl. Phys. Lett. 80 (2002) 2029–2031.
[36] E.B. Chancellor, J.P. Wikswo, F. Baudenbacher, M. Radparvar, D. Osterman, Heat conduction calorimeter for massively parallel high throughput measurements with picoliter sample volumes, Appl. Phys. Lett. 85 (2004) 2408–2410.
[37] F. Buchholz, A. Wolf, J. Lerchner, F. Mertens, H. Harms, T. Maskow, Chip calorimetry for fast and reliable evaluation of bactericidal and bacteriostatic treatments of biofilms, Antimicrobal Agents and Chemotherapy 54 (2010) 312–319.

CHAPTER

Nanomorphic cell communication unit

5

CHAPTER OUTLINE

5.1 INTRODUCTION

Communication is an essential function of autonomous microsystems. Suppose that the nanomorphic cell collects data that need to be communicated to a receiver external to the body. It follows that the cell must be equipped to transmit and receive data and, of course, there are space and energy questions that arise in this connection. The purpose of this chapter is to explore the physical limits for a micron-scale communication system and especially to obtain lower bound estimates for the energy required to transmit one bit of data.

For autonomous operation, the communication channel should be wireless since it is assumed that the nanomorphic cell is untethered. Also, from a practical point of view, it would be advantageous for transmission from the nanomorphic cell to be *omnidirectional*, i.e. uniform in all directions independent of the position and orientation. The limits of such convenience-based *ubiquitous* communication will be the focus of this chapter.

An important notion is that communication is the transmission of physical effects from one system to another [1], and therefore the limiting cases can be analyzed using the principles of fundamental physics, similar to those used in Chapter 3. The principal physical effect used for ubiquitous communication is *electromagnetic radiation*. The analyses offered in this chapter center on the electromagnetic transducer (e.g. the antenna) since its scaling directly impacts communication system performance.

5.2 ELECTROMAGNETIC RADIATION

Time varying electrical or magnetic fields produce self-propagating waves of electromagnetic (EM) energy, which propagate in space at the speed of light, c (in vacuum, $c = 3\times10^8$ m/s), and are used to transfer energy and information. Specifically, if an electron moves with *acceleration* (or deceleration), it produces electromagnetic radiation. When an alternating current (AC) of frequency ν is flowing in a conductor, electromagnetic radiation of the same frequency is emitted from the conductor into its external environment. The generated electromagnetic wave propagates through space and it can affect the behavior of other electrons at large distances from the emitter. This constitutes the basic principle of 'wireless' communication, e.g. radio.

Electromagnetic waves are, as all other waves, characterized by their frequency, ν, and wavelength, λ (see Box 5.1). Electromagnetic radiation is classified according to the frequency/wavelength, as shown in Figure 5.1. For wireless communication, radio, microwave and infrared parts of the spectrum are typically used. In the case of nanomorphic cell, radiation wavelengths in the THz to PHz regime ($\lambda \sim 10$–100 μm) are comparable to the physical size of the cell.

While in classical electrodynamics and in most engineering treatments electromagnetic radiation is regarded as a continuous phenomenon, quantum physics has revealed the granular nature of electromagnetic radiation, which is primarily manifested through the interaction of radiation with matter, e.g. in emission and absorption processes. The smallest amount of radiation energy, which can be emitted or absorbed, is called the quantum of radiation or photon (Box 5.2). The quantum nature of the electromagnetic radiation becomes especially apparent when the scale of objects and interactions is small, e.g., in the micrometer and nanometer range – the nanomorphic cell.

Although an electromagnetic signal can be physically analyzed in various ways, it will be instructive to consider extreme cases only. One extreme is a classical continuous-wave detection, which implies a large number of photons. This case typically occurs in radiofrequency (RF) communication. The other detection extreme is a 'counter', an instrument which records the incidence of single photons. The discrete photon regime is more typical for optical communication. In the following the basic principles of both RF and optical communication will be considered.

5.3 BASIC RF COMMUNICATION SYSTEM

A schematic block diagram for a radio communication system is shown in Figure 5.1. It consists of a *transmitter* and a *receiver*. The essential components of a radio transmitter are:

T1: An oscillator that generates AC current of certain carrier frequency, which is the source of EM radiation.

T2: An antenna, which serves as a transducer to convert the AC current into EM radiation.
T3: A power amplifier that increases the strength of the AC signal fed to the antenna (optional).
T4: A modulator that transfers the user's information to the EM carrier.

The radio receiver typically consists of:

R1: An antenna, the sensor/transducer, that absorbs the EM radiation and converts it into AC current.
R2: The 'tuner' selects the AC frequency of a given carrier from a multitude of frequencies absorbed by the antenna.
R3: An amplifier to increase the strength of the AC signal received by the antenna (optional).
R4: A demodulator (detector) to extract the original informational signal from the AC carrier.

BOX 5.1 TRAVELING WAVE

Wave propagation produces periodic features both in time and in space as shown in Figure B5.1 for a sine wave. The time component is characterized by the oscillation period Θ (the time for one complete cycle of an oscillation, e.g., the time interval between two neighbor maxima points **1** and **2**. $\Theta = t_2 - t_1$ in Fig. B5.1) and frequency $\nu = 1/\Theta$ (the number of the cycles per unit time), as shown in Figure B5.1. In turn, the distance between two neighbor wave maxima in space (say between points **1*** and **2***) is called the wavelength. In a traveling wave, the relationship between the time and space components can be derived from the basic motion relationship: $L = ut$. Thus if $u = c$ (the electromagnetic wave in vacuum) and $t = \Theta$, the distance between the corresponding maxima points in space, **1*** and **2***, that the wave propagates during one period is

$$\lambda = c \cdot \Theta = \frac{c}{\nu} \quad \text{(B5.1)}$$

A simple example of a traveling wave is the *plane wave*, which is described by the sine function

$$A(x,t) = A_0 \sin\left(2\pi\nu t - \frac{2\pi}{\lambda}x\right) \quad \text{(B5.2)}$$

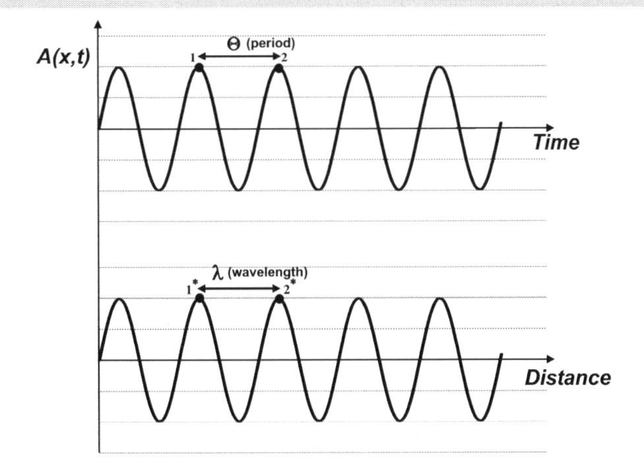

FIGURE B5.1

Space and time components of a traveling wave

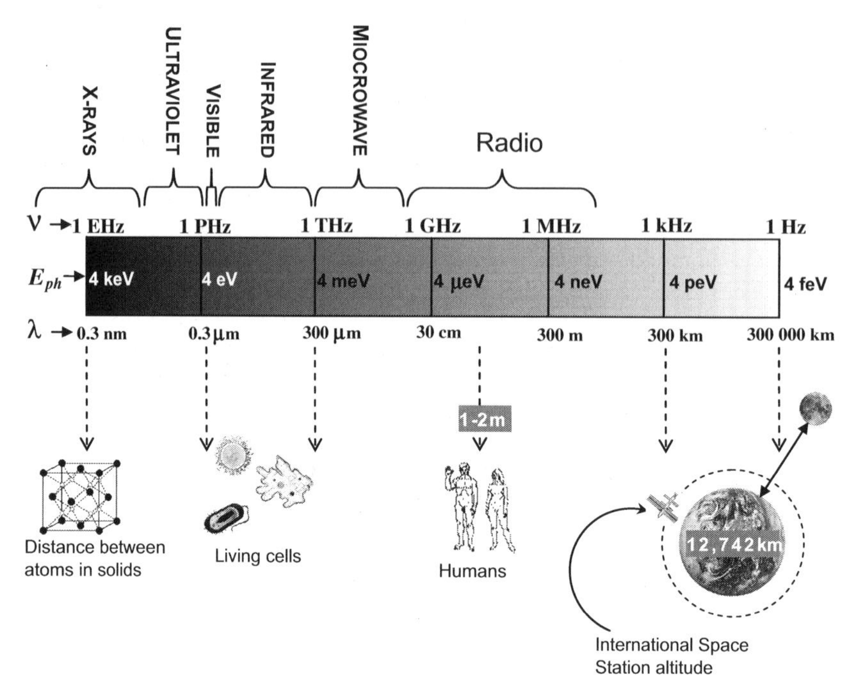

FIGURE 5.1

Electromagnetic spectrum

BOX 5.2 PLANCK-EINSTEIN EQUATION

In 1905, the general validity of the concept of continuous electromagnetic radiation was seriously called into question in an article by Albert Einstein(1). In this work, Einstein argued that light and other forms of electromagnetic radiation can be emitted or absorbed only in discrete quantities, the 'radiation quanta', which were later also called *photons*. The energy of the photons, E_{ph}, was shown to be related to the frequency and wavelength of the radiation as

$$E_{ph} = h\nu = \frac{hc}{\lambda} \tag{B5.3}$$

where h is Planck's constant, $h = 6.63 \times 10^{-34}$ J·s.

Relation (B5.3), sometimes called Planck-Einstein equation, revealed the granular structure of electromagnetic radiation and it was used to explain experimental phenomena such as the photoelectric effect. In fact this was the primary discovery, for which Einstein was awarded the Nobel Prize in 1921 'for his services to Theoretical Physics, and especially for his discovery of the law of the photoelectric effect'.

The notion of the light quantum as a particle also carries with it the concept of particle 'size'. For many purposes, for example to choose a detector of appropriate width, an effective 'size' for the photon is roughly its wavelength, λ.(2)

(1) A. Einstein, "Generation and conversion of light with regard to a heuristic point of view", Ann. Phys. 17 (1905) 132–148.

(2) B.R. Frieden, "Information-based uncertainty for a photon", Optics Commun. 271 (2007) 71–72.

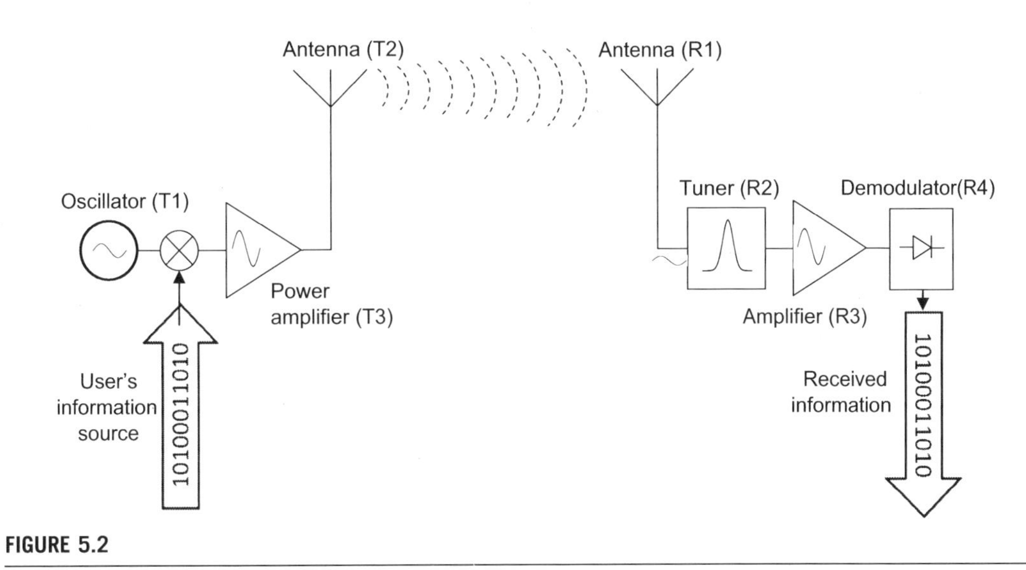

FIGURE 5.2

Block diagram of a radio communication channel

5.4 EM TRANSDUCER: A LINEAR ANTENNA

5.4.1 Basic principles

One way to produce EM radiation is to create a directional flow of *accelerating/decelerating* electrons, e.g., by inducing alternate electrical current in a linear metal wire – an antenna, as shown in Figure 5.3. The radiation properties of such an antenna strongly depend on the antenna length. The effects of the antenna size are due to the propagation properties of the wave along the antenna wire. When an oscillating voltage source is attached to one end of the antenna, the electric field will push the first electron (Fig. 5.3), which will in turn push the second one, then the third one, etc., thus creating a train of electron displacements. The speed of the train is close to the speed of light, c. These displacements will propagate until the other end of the wire is reached, at which point the last electron hits a barrier and is reflected back, changing the direction of its movement to an opposite (in this idealistic picture, possible energy losses are neglected). Now there are two trains, moving in opposite directions and if they 'collide', wave propagation is annihilated. Thus, efficient radiation by a short antenna is possible only during the time interval, τ, before the wave propagates distance L_{ant} to the end of the antenna:

$$\tau = \frac{L_{ant}}{c} = \frac{L_{ant}}{\lambda}\Theta \tag{5.1}$$

For longer antennas, conditions can be found for non-destructive wave propagation along the antenna. Destructive 'collisions' of waves are avoided when the direct and the reflected waves are synchronized. This means that the wave must arrive to its point of origin after traveling distance $2L_{ant}$ in the same phase, i.e. the time to travel is equal to the period of the oscillation:

$$\frac{2L_{ant}}{c} = n\Theta \tag{5.2a}$$

or

$$\frac{2L_{ant}\Theta}{\lambda} = n\Theta \tag{5.2b}$$

where n is an integer ($n = 1, 2, \ldots$). Thus an 'optimum' antenna length is

$$L_{opt} = n\frac{\lambda}{2} \tag{5.2c}$$

Common practical antennas are the half-wave antenna ($n = 1$, $L_{ant} = \lambda/2$) and the full-wave antenna ($n = 2$, $L_{ant} = \lambda$).

5.4.2 Short antennas

While short antennas ($L_{ant} << \lambda$) are much less efficient, they are often used when the system dimensions are strictly constrained. Below an intuitive order-of-magnitude analysis of the radiation by a short antenna is offered. A short straight wire has a capacitance C_{wire}:

$$C_{wire} \sim \varepsilon_0 L_{ant} \tag{5.3}$$

The corresponding energy stored in the wire is

$$E = \frac{CV^2}{2} \tag{5.4}$$

For simplicity, the alternating (e.g. sine form) voltage $V(t)$ applied to the wire will be approximated by triangle wave, as shown in Figure 5.3:

$$V_\Delta \sim 4V_0\frac{t}{\Theta} = 4V_0\frac{L_{ant}}{\lambda} \tag{5.5}$$

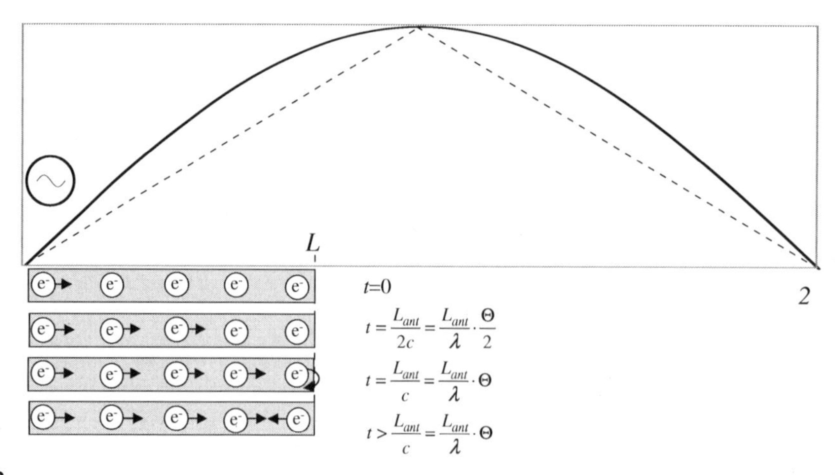

FIGURE 5.3

An intuitive interpretation of wave propagation and energy loss in short antennas

The average value of voltage in the interval from 0 to L_{ant} ($L_{ant} << \lambda$) is

$$\langle V_\Delta \rangle = 2V_0 \frac{L_{ant}}{\lambda} \tag{5.6}$$

Putting (5.6) into (5.4) obtain:

$$E = \frac{C\langle V_\Delta \rangle^2}{2} \sim \frac{1}{2}\varepsilon_0 L_{ant} \cdot 4V_0^2 \left(\frac{L_{ant}}{\lambda}\right)^2 = 2\varepsilon_0 L_{ant} V_0^2 \left(\frac{L_{ant}}{\lambda}\right)^2 \tag{5.7}$$

Next, the maximum power which can be radiated during the time interval, τ, of non-destructive wave propagation (5.1) is

$$P_{rad} = \frac{E}{\tau} = E \cdot \frac{c}{L_{ant}} \tag{5.8}$$

Substituting (5.7) into (5.8), obtain the maximum power radiated by a short antenna:

$$P_{rad} \sim 2\varepsilon_0 L_{ant} V_0^2 \left(\frac{L_{ant}}{\lambda}\right)^2 \cdot \frac{c}{L_{ant}} = 2 \cdot V_0^2 (\varepsilon_0 c) \cdot \left(\frac{L_{ant}}{\lambda}\right)^2 \sim \frac{V_0^2}{Z_0}\left(\frac{L_{ant}}{\lambda}\right)^2 \sim I_0^2 Z_0 \left(\frac{L_{ant}}{\lambda}\right)^2 \tag{5.9}$$

where $Z_0 = \frac{1}{\varepsilon_0 c} \approx 377\ \Omega$ is the impedance of free space.

Note that $Z_0 \left(\frac{L_{ant}}{\lambda}\right)^2$ in (5.9) has dimensions of resistance and is called the radiation resistance of the antenna:

$$R_{rad} \sim Z_0 \left(\frac{L_{ant}}{\lambda}\right)^2 \sim 377 \left(\frac{L_{ant}}{\lambda}\right)^2 \tag{5.10a}$$

More rigorous derivations of the short antenna radiation resistance can be found in the antenna theory literature [2,3]. The resulting expressions can differ depending on the input assumptions. For example, Blake [2] obtains (assuming uniform current distribution):

$$R_{rad} \approx 790 \left(\frac{L_{ant}}{\lambda}\right)^2 \tag{5.10b}$$

while Jackson [3] gives (for non-uniform current distribution):

$$R_{rad} \approx 197 \left(\frac{L_{ant}}{\lambda}\right)^2 \tag{5.10c}$$

The formulas (5.10a–10c) were derived for short antenna approximations assuming $L_{ant} << \lambda$ and, generally speaking, they are not applicable to the longer antennas with $L_{ant} \sim \lambda$. For example, an exact solution for an ideal half-wave antenna is [2]

$$R_{\lambda/2} = 73.1\Omega \tag{5.10d}$$

It is interesting to note that application of (5.10a) to a half-wave antenna yields an estimate of 94 Ω in radiation resistance, which is remarkably close to the value given in (5.10d). This suggests that the functional form of (5.10a) can be used as an order-of-magnitude approximation up to $L_{ant} = \lambda/2$.

5.4.3 Radiation efficiency

According to (5.9), the radiated EM power, P_{rad}, in short antennas is diminished by the square of the ratio of antenna length to wavelength. The total input power, P_{in}, needed to radiate the power P_{rad} is the sum of the radiated power and loss power (conductor and dielectric loss):

$$P_{in} = P_{rad} + P_{loss} \tag{5.11}$$

The antenna radiation efficiency can be defined as

$$\eta_{rad} = \frac{P_{rad}}{P_{in}} = \frac{P_{rad}}{P_{rad} + P_{loss}} \tag{5.12}$$

The antenna circuit can be represented as a loss resistance R_{loss} in series with radiation resistance R_{rad} (Fig. 5.4). The corresponding radiated and loss powers are:

$$P_{rad} = I^2 R_{rad} \tag{5.13a}$$

$$P_{loss} \sim I^2 R_{loss} \tag{5.13b}$$

From (5.13) and (5.10) obtain:

$$\frac{P_{rad}}{P_{loss}} \sim \frac{Z_0}{R_{loss}} \left(\frac{L_{ant}}{\lambda} \right)^2 \tag{5.14}$$

$$\eta_{rad} = \frac{P_{rad}}{P_{rad} + P_{loss}} = \frac{(P_{rad}/P_{loss})}{(P_{rad}/P_{loss}) + 1} = \frac{\frac{Z_0}{R_{loss}} \left(\frac{L_{ant}}{\lambda} \right)^2}{\frac{Z_0}{R_{loss}} \left(\frac{L_{ant}}{\lambda} \right)^2 + 1} \tag{5.15}$$

Since Z_0/R_{loss} is a constant for a particular configuration, it follows that for sufficiently small L_{ant}, the radiation efficiency is approximately

$$\eta_{rad} \approx \frac{Z_0}{R_{loss}} \left(\frac{L_{ant}}{\lambda} \right)^2 \tag{5.16a}$$

Next, as it is argued in [4], high ohmic losses are expected for very small antennas operating in the THz regime. Even if an ideal antenna with resistivity of bulk copper is considered, the resistance of a wire 50 μm long and 0.1 μm in diameter would be more than 100 Ω. Additionally, in wireless transmission of information from the human body, the antenna is immersed in an electrically lossy medium which will further increase R_{loss} [5]. It can therefore be assumed in the order-of-magnitude estimates of nanomorphic cell antenna that $R_{loss} \sim Z_0$ (377 Ω), and thus

$$\eta_{rad} \sim \left(\frac{L_{ant}}{\lambda} \right)^2 \tag{5.16b}$$

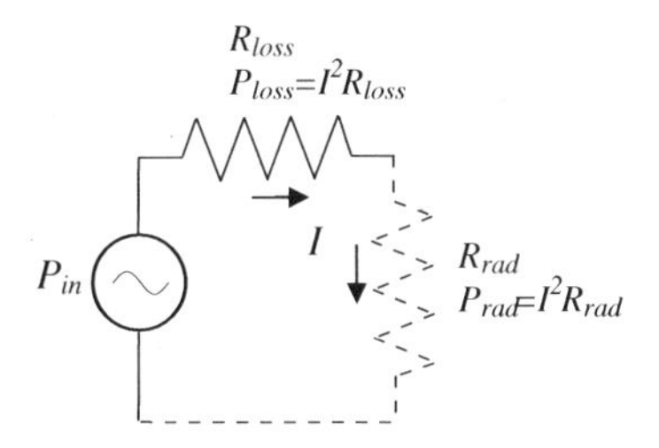

FIGURE 5.4

Depiction of a typical antenna circuit model

BOX 5.3 ANTENNA

An antenna system provides a practical means of transmitting to a distant point in space energy (in the EM form) and information. The antenna performance is characterized by the efficiency of transmission and the signal distortion.[1] In general, for an efficient antenna the antenna length should be comparable to the operational wavelength, $L_{ant} \sim \lambda$.

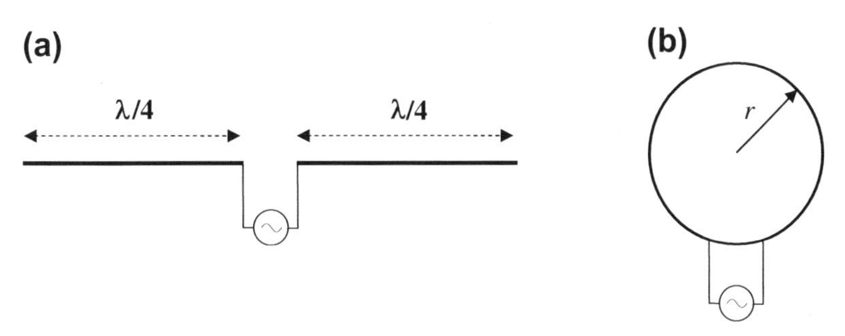

FIGURE B5.3

Common simple antennas: (a) linear half-wave dipole and (b) loop antenna

Some common simple antennas are the linear half-wave dipole (Fig. B5.3a) and the loop antenna (Fig. B5.3b). Small loop antennas (i.e. the loop circumference $2\pi r < \lambda/4$) are often used in size-constrained applications.

(1) L.J. Chu, "Physical limitations of omni-directional antennas", J. Appl. Phys. 19 (1948) 1163–1175.

5.5 FREE-SPACE SINGLE-PHOTON LIMIT FOR ENERGY IN EM COMMUNICATION

First, a communication system operation in free space will be considered so that the effects of the transmission medium such as absorption of radiation can be ignored. The minimum energy requirements in EM communication can be estimated based on the fact that the radiation can be emitted or absorbed only in discrete increments – photons (see Box. 5.2). This means, among other things, that at least one photon must be absorbed by a receiving device and the photon energy (Box 5.2) can be viewed as the lower bound on the energy of EM communication.

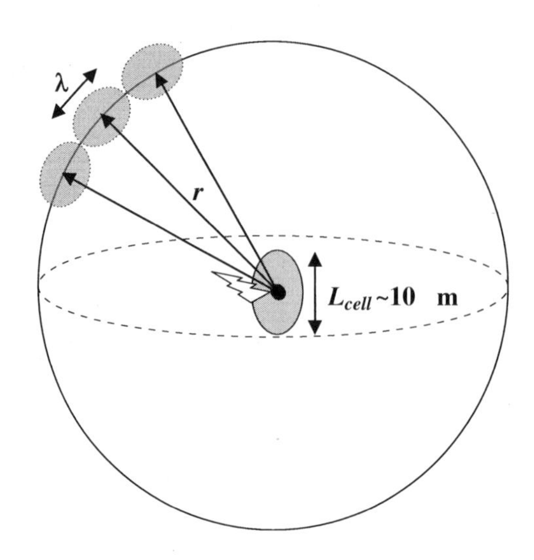

FIGURE 5.5

Illustration of communication between a nanomorphic cell and an external device

Consider the case of a uniformly radiated electromagnetic wave connecting a 10-μm nanomorphic cell to an external device located at a distance, *r*, from the cell (Fig. 5.5). As was stated above in the limit, at least one transmitted photon must be absorbed by the external detector for successful communication. If the location of the external receiving device relative to the nanomorphic cell is unknown, then in order to guarantee that at least one photon will reach the detector, the entire sphere of radius *r* must be 'covered' with photons. The area 'covered' by one photon of wavelength λ is $\sim\lambda^2$ and therefore the total number of photons needed to be emitted into the solid angle of 4π steradians (a complete sphere) is

$$N_{4\pi} \sim \frac{4\pi r^2}{\lambda^2} \tag{5.17}$$

Note that (5.17) is a variation of the Friis formula for signal strength attenuation with increased distance [6,7].

The energy of each photon is

$$E_{ph} = h\nu = \frac{hc}{\lambda} \tag{5.18}$$

Thus, the energy associated with one full 'communication packet', i.e. the minimum energy required to transmit one bit of information such that it is equally accessible at all points on the sphere, is approximately

$$E_{com} = N_{4\pi} \bullet E_{ph} \sim \frac{4\pi r^2}{\lambda^2} \bullet \frac{hc}{\lambda} = \frac{4\pi hcr^2}{\lambda^3} \tag{5.19}$$

As was discussed in the previous section, the size of the transducer (e.g. the antenna) needs to be about the same as the radiated wavelength, λ (in order to maximize antenna efficiency). Let the antenna size be limited by the cell size, i.e. $\lambda \sim L_{cell} \sim 10$ µm. If the distance between the cell and the receiver $r = 1$ m, Eq. (5.19) gives $E_{com} \sim 2.5\times10^{-9}$ J/bit. Note that this energy estimate is a lower bound on communication and it does not consider, e.g., efficiencies of the transducer and detector, noise, etc.

The above result (5.19) reveals that the ubiquitous (i.e. omnidirectional) communication by the extreme microsystem is relatively costly from an energy point-of-view. For example, given a total available energy of $\sim 10^{-5}$ J, as was evaluated in Chapter 2, and $E_{com} \sim 2.5\times10^{-9}$ J/bit, the maximum number of bits the cell could send is

$$N_{bit} = \frac{E_{stored}}{E_{com}} \sim \frac{10^{-5} J}{2.5 \times 10^{-9} J/bit} \sim 4000 \text{ bits} \tag{5.20}$$

A comparison between the number of binary switching operations N_{bit} obtained with 10^{-5} J (see Chapter 3) and the number of omnidirectional bits transmitted with the same amount of energy is given below.

Computation	Communication
$E_{SW} \sim 10^{-18}$ J	$E_{com} \sim 10^{-9}$ J
$N_{bit} \sim 10^{13}$	$N_{bit} \sim 10^{3}$

As can be seen, the communication is a very costly process compared to computation. The undirected transmission of one bit of data is vastly more costly from an energy point of view than switching the state of a single transistor. Note that the above derivation is very conservative since neither issues of *noise* nor *transmission losses* have been taken into account. This will be considered in the following sections.

It is instructive to compare the result (5.19) to the existing radios under consideration for the wireless sensor network applications [7,8]. Figure 5.6 displays the experimental data along with the

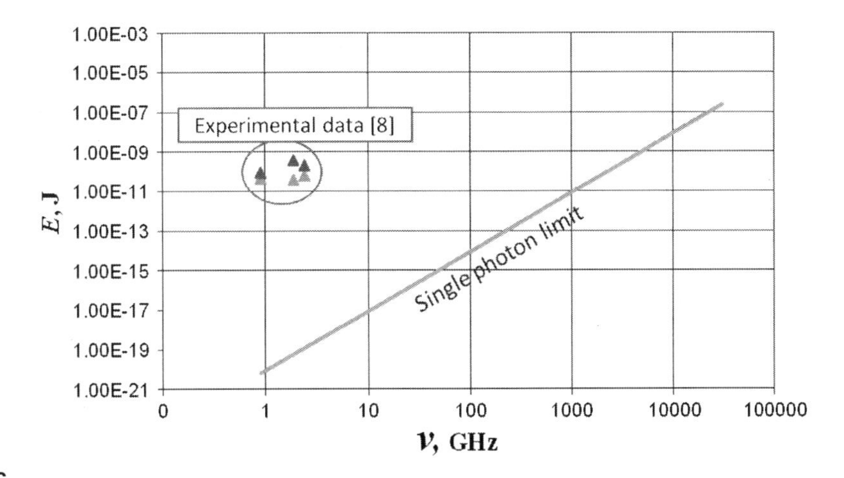

FIGURE 5.6

Energy to send a bit: a comparison of the single photon limit (5.13) with the energy per 'useful bit' data for wireless sensor network communication [8]

single photon limit (5.19). It can be seen that in the 1 GHz frequency range (~cm system size) the energy per 'useful bit' (defined in [8] as a weighted average energy) is $\sim 10^{-11}$–10^{-10} J for communication distance $r = 10$ m. Note that the experimental data are ten orders of magnitude above the corresponding single-photon limit (Fig. 5.6). This suggests there might be ways to reduce the energy per communicated bit. Indeed, a number of schemes have been proposed to minimize energy use in communication, e.g., by data rate optimization, simpler modulation schemes, etc. [7,8]. As can be seen from Figure 5.6, however, as smaller system sizes are considered and therefore higher carrier frequencies are used, the improvement margin decreases. Therefore it will be difficult to lower the communication energy beyond the current levels. For a 10-μm-scale system size, even the single-photon energy limit is considerably larger than the practical communication energy for the cm-scale systems.

Several options could be considered to minimize the energy expenditures for communication. First, as follows from (5.19), $E_{com} \sim 1/\lambda^3$ and therefore using EM radiation of longer wavelength appears to be beneficial as it would reduce communication energy. However, as was discussed in the previous section, efficient radiation requires the antenna size to be comparable with the wavelength (e.g. $\lambda/2$). One option is to use an external antenna whose length is larger than the size of the nanomorphic cell. Carbon nanotubes are promising for such antennas [4,9] as they offer suitable length range, smaller diameter, and good conductive properties. Another problem arising from the use of longer wavelengths is the decreasing energy of photons (5.18) and thus a larger number of photons is needed to overcome thermal noise. The option of using longer wavelength radiation for nanomorphic cell communication will be discussed below.

5.6 THERMAL NOISE LIMIT ON COMMUNICATION SPECTRUM

5.6.1 Thermal background radiation

Derivations of energy for communication in the previous section assume a limiting case when only one transmitted photon is absorbed by the external detector to generate a distinguishable signal. This assumption can be regarded as a plausible theoretical limit, if the photon energy is larger than the energy of thermal photons, i.e. $E_{ph} >> k_BT$. However, for $\lambda \sim L_{cell} \sim 10\ldots50$ μm this exactly coincides with the peak of thermal radiation at $T = 300$ K (see Fig. 5.7). Thus, communication in this range is not a feasible option due to the very strong background noise. Therefore it is necessary to explore communication options for the nanomorphic cell beyond the thermal radiation band (circa 5–50 nm). One option is to use larger wavelength, e.g. $\lambda > 100$ μm (THz range); the other option is to go a much shorter wavelength, e.g. $\lambda \sim 1$ μm (near infrared optical range).

5.6.2 Minimum detectable energy

Next, as it follows from the physics of radiation (see Box 5.2), the minimum energy required for the transmission of one bit of information cannot be less than one quantum, $h\nu$. On the other hand, no communication is possible with an energy less than k_BT. The minimum detectable energy due to the $h\nu$ and k_BT constraints is shown in Figure 5.8. In the shorter wavelength region of Figure 5.8, where $h\nu > k_BT$, the smallest detectable energy is that of one photon which must have energy greater than k_BT. As the wavelength is increased, the energy of the photon will decrease until it becomes smaller than k_BT.

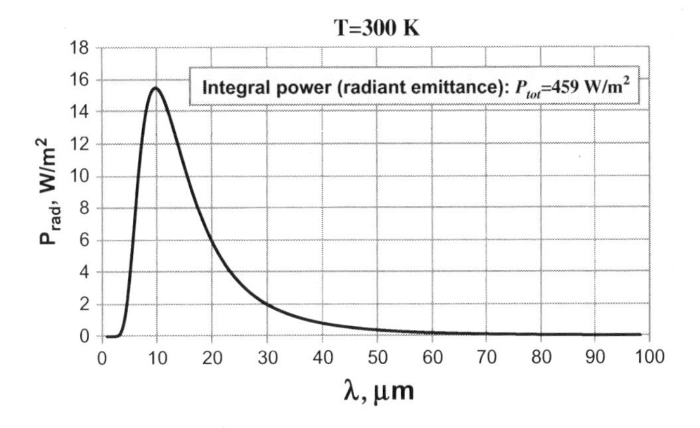

FIGURE 5.7

Thermal radiation spectrum at $T = 300$ K

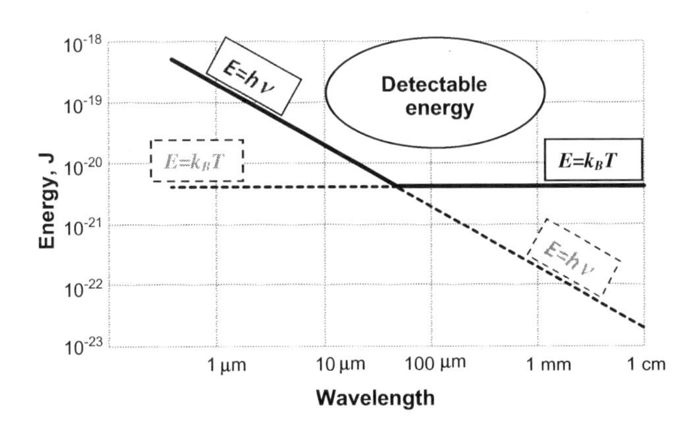

FIGURE 5.8

Minimum detectable energy at $T = 300$ K

(adapted from [10])

In this case, single-photon detection is impossible and many photons need to be received for each one-bit package:

$$N_{ph} > \frac{k_B T}{h\nu} = \frac{k_B T}{hc}\lambda \tag{5.21a}$$

If one assumes the minimum detectable energy to be a factor of $x \geq 1$ larger than the thermal energy $k_B T$, the minimum number of photons in the 'detectable bit' is

$$N_{ph}^{\min} = \frac{x k_B T}{h\nu} = \frac{x k_B T}{hc}\lambda \tag{5.21b}$$

5.7 THE THz COMMUNICATION OPTION (λ ≥ 100 μM)

As was discussed in Section 5.4, using EM radiation of longer wavelength appears to be beneficial as it would reduce communication energy. This would require an external antenna with a length larger than the size of the nanomorphic cell. An example of a nanomorphic cell with an external half-wave antenna for 100-μm transmission is shown in Figure 5.9 ('to scale'). This corresponds to 3 THz frequency and such a range is currently unattainable by existing technologies (see Box 5.4).

According to (5.21a) and Figure 5.8, communication at $\lambda = 100$ μm occurs in a multiphoton mode and the communication energy per bit can be estimated by multiplying (5.19) by (5.21b):

$$E_{com} \sim \frac{4\pi hcr^2}{\lambda^3} \cdot \frac{xk_BT}{hc}\lambda = \frac{x4\pi r^2}{\lambda^2} k_BT \tag{5.22}$$

The numerical result for $\lambda = 100$ μm and $r = 1$ m gives $E_{com} \sim x \cdot (5 \times 10^{-12})$ J.

Next, the number of photons needed can be estimated based on the power signal-to-noise ratio argument discussed in Chapter 4. The detection limit in the presence of noise is set by the signal-to-noise ratio (SNR), which is usually defined in terms of a power ratio. A 'minimal' requirement on the SNR can be written as

$$SNR = \frac{P_s}{P_n} = 2 \tag{5.23}$$

When the detector absorbs one photon of energy $E_{ph} = h\nu$, the corresponding signal power is:

$$P_s = P_{ph} \sim \frac{E_{ph}}{t} \tag{5.24}$$

where $t = t_H$ is Heisenberg time, defined in Chapter 3:

$$t_H = \frac{h}{2E_{ph}} \tag{5.25}$$

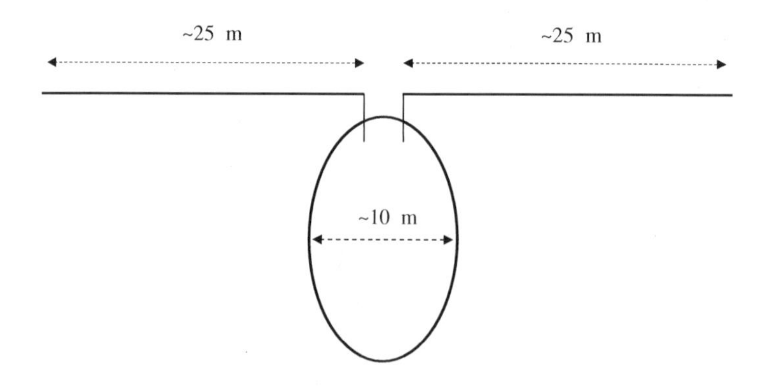

FIGURE 5.9

A nanomorphic cell with an external half-wave antenna for 100 μm

BOX 5.4 T-RAYS: CHALLENGES AND OPPORTUNITIES

Terahertz EM radiation (T-rays) with frequencies ~100 GHz–10 THz (free-space wavelength ~30 μm–3 mm) lies at the boundary between light and radio waves. The terahertz regime is one of the least explored spectral bands and is currently regarded by some as very promising for a growing range of applications, including screening, biomedical imaging, and wireless communication.[1–4]

There are two difficult problems in practical use of the T-rays:

1. Transmission THz gap: The THz radiation is strongly absorbed by water and water vapor. This makes it difficult to use THz radiation for communication in vivo.
2. Instrumental THz gap: There are currently very limited device capabilities for signal generation and detection in this frequency range.

While sensitivity of the terahertz radiation to water is a significant disadvantage for communication, this may turn out to be very beneficial for the purposes of bio-sensing and bio-imaging, as it allows one, e.g., to quantify the water content of biological tissues.[4] The 'instrumental THz gap' is being gradually narrowed by advances in silicon CMOS FET technology, which has already entered the sub-THz domain (>100 GHz). New devices are also being investigated, e.g., the resonant tunneling diode (RTD) is a promising device for THz generation.[5] Ideas for using carbon nanotubes for terahertz emitters and detectors have even been recently discussed.[6]

(1) J. Grade, P. Haydon, and D. van der Weide, "Electronic terahertz antennas and probes for spectroscopic detection and diagnostics", Proc. IEEE 95 (2007) 1583–1591.
(2) I. Hosako, N. Sekine, M. Patrashin, S. Saito, K. Fukunaga, Y. Kasai, P. Baron, T. Seta, J. Mendrok, S. Ochiai, and H. Yasuda, "At the dawn of a new era in terahertz technology", Proc. IEEE 95 (2007) 1611–1623.
(3) A.R. Orlando and G.P. Gallerano, "Terahertz radiation effects and biological applications", J. Inrared Milli. Terahz Waves 30 (2009) 1308–1318.
(4) W.L. Chan, J. Deibel, and D.M. Mittleman, "Imaging with terahertz radiation", Rep. Progr. Phys. 70 (2007) 1325–1379.
(5) K. Urayama, S. Aoki, S. Suzuki, M. Asada, H. Sugiyama, and H. Yokoyama, "Sub-terahertz resonant tunneling diode oscillators integrated with tapered slot antennas for horizontal radiation", Appl. Phys. Express 2 (2009) 044501.
(6) M. Rosenau da Costa, O. S. Kibis, M.E. Portnoi, "Carbon nanotubes as a basis for terahertz emitters and detectors, Microelectronics J. 40 (2009) 776–778.

BOX 5.5 TRANSMISSION AND ABSORPTION OF EM RADIATION

When an EM radiation beam passes through a physical medium different from a vacuum, part of its energy is absorbed by the medium. As a result, the intensity (power density) of the beam transmitted through a layer of substance with thickness Electromagnetic spectrum δ is reduced according to the *Beer-Lambert* law:

$$p = p_0 \exp(-\alpha \cdot \delta) \quad \text{(BV5.1)}$$

where p_0 and p are the power densities (intensities) respectively of the incident radiation and radiation coming out and α is the absorption coefficient, which is characteristic of a given substance.

The related parameter characterizing the transparency of a medium to the radiation is the transmission coefficient, η:

$$\eta = \frac{p}{p_0} \quad \text{(BV5.2)}$$

From (BV5.1) and (BV5.2):

$$\eta = \exp(-\alpha \cdot \delta) \quad \text{(BV5.3)}$$

From (5.24) and (5.25), the single-photon power is

$$P_s \sim \frac{E_{ph}}{t_H} \sim \frac{2E_{ph}^2}{h} = \frac{2hc^2}{\lambda^2} \tag{5.26}$$

Next, the thermal noise power derived in Chapter 4 is

$$P_n = 4k_B T \Delta \nu \tag{5.27}$$

The bandwidth $\Delta\nu$ is related to the signal's time duration $t_s = t_H$ as

$$\Delta\nu \sim \frac{1}{t_s} \tag{5.28}$$

From (5.27) and (5.28):

$$P_n = 4k_B T \frac{2E_{ph}}{h} = 8k_B T \frac{c}{\lambda} \tag{5.29}$$

Now, the minimum number of photons per one-bit signal, N_{ph}^{min}, can be obtained combining (5.23), (5.26), and (5.29):

$$\frac{P_n}{P_s} = N_{ph}^{min} \frac{2hc^2}{\lambda^2} \bullet \frac{1}{8k_B T} \bullet \frac{\lambda}{c} = N \frac{hc}{\lambda} \frac{1}{4k_B T} = 2 \tag{5.30a}$$

$$N_{ph}^{min} = \frac{8k_B T}{hc} \lambda \tag{5.30b}$$

If one now compares (5.30b) and (5.21b), the factor x in (5.21b) can be found: $x = 8$. By substituting $x = 8$ into (5.22), the lower bound on the energy per bit in longer wavelength communication ($\lambda >> L_{cell}$) is

$$E_{com} \sim \frac{32\pi r^2}{\lambda^2} k_B T \tag{5.31}$$

The numerical result for $\lambda = 100$ μm and $r = 1$ m gives $E_{com} \sim 4 \times 10^{-11}$ J.

While the estimate of communication energy per bit for $\lambda = 100$ μm is encouraging, since it is almost two orders of magnitude smaller than the estimate in Section 5.5, additional practical constraints need to be taken into account. The most important constraint is transmission losses. All living tissues contain a large amount of water; therefore, the absorption of EM radiation by water cannot be ignored for in vivo operations. In fact, the maximum absorbance of EM radiation by water is between 100 and 10 μm, as can be seen in the absorption spectrum in Figure 5.10a. Figure 10b shows the corresponding transmission spectrum of THz radiation through a layer of water 100 μm thick. For $\lambda = 100$ μm, the transmission coefficient is close to 10^{-3}, which raises the communication energy per bit 1000×to $\sim 10^{-8}$ J. If one assumes acceptable transmission loss to be, for example, somewhat less then 10×, then according to Figure 5.10b, the wavelength should be ~1 mm or larger. If the wavelength is increased without a corresponding increase in antenna length, the

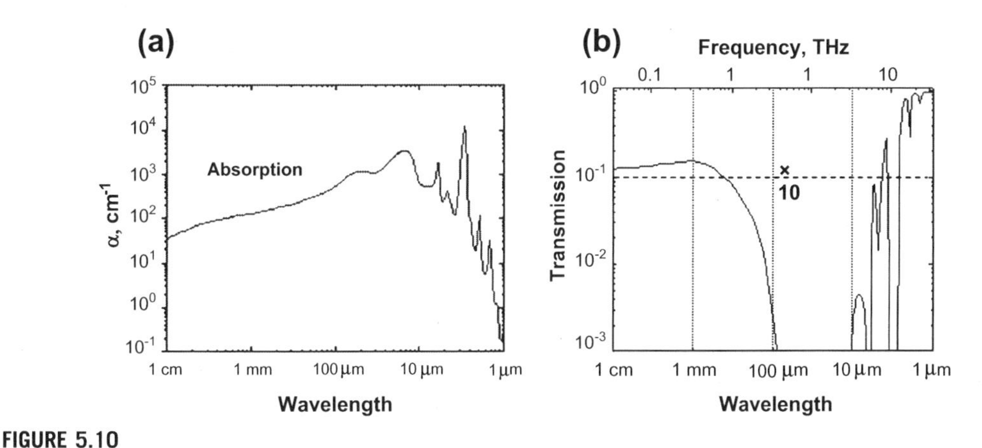

FIGURE 5.10

Absorption (a) and transmission (b) spectra of THz and IR radiation in water [11]

radiation efficiency will be diminished by the square of the ratio of antenna length to wavelength, as it was discussed in Section 5.4:

$$E_{com} \sim \frac{32\pi r^2}{\lambda^2} k_B T \left(\frac{\lambda}{L_{ant}}\right)^2 = \frac{32\pi r^2}{L^2} k_B T \qquad (5.32)$$

The numerical result for $L_{ant} = 50$ µm and $r = 1$ m gives $E_{com} \sim 2 \times 10^{-10}$ J. This number represents an optimistic estimate as it does not include many other possible noise sources, first of all the photon shot noise [1,10], which makes the single-photon communication problematic.

5.8 WIRELESS COMMUNICATION FOR BIOMEDICAL APPLICATIONS

Ultra-small wireless communication systems are an important part of emerging bioelectronic devices, including wearable and implantable devices. These devices may further be organized in wireless body networks (BAN) [12]. The need for dramatic miniaturization drives the need for revision of the traditional design of communication systems such as shown in Figure 5.2 [12]. For example, in the transmitter, a loop antenna could be used as an element of the oscillator that defines the carrier frequency, thus making a power amplifier unnecessary for the short transmission range [12]. Progress in microelectronic device technology allows for further scaling of most of the circuit components of the communication system, with the antenna being a remarkable exception. In fact, a key challenge in the design of communication systems for on- or in-body application is the stringent constraint on antenna size [12], which is a critical issue [12–15]. One example of an application where extremely small antennas are needed is the retinal prosthesis system [13]. Various techniques are used to make miniaturized antennas, such as meander line, spiral geometries, fractal designs, etc. Recently, 3D wire dipole antennas have been shown to be promising for implantable data-telemetry in retinal prosthesis [13].

The antenna issue has been made even more difficult for today's practical bioelectronic devices because of existing regulations. The RF frequencies allocated by current regulations for use in

Table 5.1 Capsule endoscopy image transmitter systems

Group	Power, mW	Frequency, MHz	Energy/bit (nJ)	Transmitter efficiency
Zarlink ZL70101* [16]	17.5	403/434	21.9	2.03%
Nordic nRF2401* [16]	37.8	2400/2500	37.8	2.65%
AnSem NV, 2004 [17]	1.8	433	18.0	5.56%
Natl. Tsing Hua U, 2005 [18]	4	416	2	0.1%
Tsinhua U, 2007 [19]	7.9	2400	7.9	0.06%
Natl. Taiwan U, 2007 [20]	19.5	400	13.0	0.65%
MIT, 2007 [21]	6	1350–1750	4.9	20.57%
KU Leuven, 2009 [16]	2	144	1	0.79

**Commercial product.*
(adapted from [16])

biomedical implants are in the Medical Device Radiocommunication Service band (MedRadio) at 401–406 MHz. The corresponding wavelength is more than 1 m, which makes the efficient $\lambda/2$ antennas impractical [14], transmission inefficient, and thus power consumption by transmitters relatively high. Table 5.1 shows examples of data for image transmitter systems used in swallowable capsule endoscopes [16].

5.9 OPTICAL WAVELENGTH COMMUNICATION OPTION ($\lambda \sim 1\ \mu$M)

As was indicated in Section 5.6.1, communication at $\sim 1\ \mu$m (infrared radiation) is another option to avoid the thermal radiation peak (Fig. 5.7). This wavelength also has good transmission properties through water (Fig. 5.10). On the other hand, omnidirectional communication by the nanomorphic cell is practically forbidden, since according to (5.19) the minimum energy required to send a bit of information at $\lambda = 1\ \mu$m is $\sim 2.5 \times 10^{-6}$ J, which would be close to the total energy budget of the nanomorphic cell. In principle, directional transmission would reduce the number of photons in the packet and therefore the total energy consumed. However, the orientation problem must be solved, which may require additional energy expenditure. One not very elegant solution would be to use as many external detectors as possible or even a surround of receivers (e.g. MRI-like) scheme. This could limit the application space, but would reduce transmission energy. In general, the short wavelengths do not allow *ubiquitous communication by the nanomorphic cell*. However, in contrast to the THz regime, generation and detection of optical ($\sim 1\ \mu$m) radiation allows for the use of devices whose physical size is commensurate with the dimensions of the nanomorphic cell.

5.9.1 Basic principles of generation and detection of optical radiation

For generation and detection of the radiation in optical range, e.g. $\lambda \sim 1\ \mu$m (near infrared), semiconductor diode-type barrier structures (see Chapter 3) are commonly used, which belong to the

BOX 5.6 OPTICAL ANTENNAS

While classical metal antennas are used as transducers of EM radiation radiowave regime (e.g. $\lambda > 1$ mm), semiconductor optoelectronic devices, such as light-emitting diodes and photodiodes are used in infrared/optical regime (e.g. $\lambda < 10$ μm). Recently the concept of an *optical antenna* has been proposed that would perform the transducer function in a fashion analogous to standard radiowave/microwave antennas[1–3]. This is currently an emerging area of research and the developments of optical antennas may dramatically improve the efficiency of conventional optoelectronic devices such as light-emitting diodes, photodiodes, and photovoltaic cells. It is interesting to note that the principle of the optical antenna is apparently utilized in living matter. For example, as was briefly discussed in Chapter 1, the *phycobilisomes*, special proteins attached to the photosynthetic membranes in cyanobacteria, act as light-harvesting antennas which enhance the efficiency of the light-harvesting process.

(1) L. Novotony, "Optical Antennas: A New Technology that Can Enhance Light-Matter Interactions," The Bridge, 39(2009) 14–20.
(2) P. Bharadwaj, B. Deutsch, and L. Novotony, "Optical Antennas," Ad5.in Optics and Photonics 1 (2009) 438–483.
(3) H. Li and X. Cheng, "Optical Antennas: A boost for infrared radiation", J. Vac. Sci. Tech B 26 (2008) 2156–2159.

family of optoelectronic devices (see however Box 5.6). The light-generating transducer is called a light-emitting diode (LED) and the light detector is a photodiode (PD). In this section, basic principles and scaling properties for LED and PD are outlined. For a detailed analysis of optoelectronic device operation, readers are referred to the basic texts on semiconductor devices, e.g. [22].

The diode (Fig. 5.11a), as was discussed in Chapter 3, is a semiconductor material system with a barrier formed by built-in charges due to ionized impurity atoms (dopants).

If the impurity atoms are charged positively (eject electrons into the system), an *n-type* region is formed in the semiconductor matrix. The extra electrons provided by the ionized dopants can move freely in the material and thus contribute to electrical conductance. These unbound electrons occupy energy states in the *conduction band* of the semiconductor as shown in Figure 5.11.

If, on the other hand, the impurity atoms are charged negatively (capture electrons from the system), a *p-type* region is formed in the semiconductor matrix. The electrons captured by dopants originate from the initially neutral atoms of the semiconductor matrix. These atoms with absent electrons are charged positively and are called *electron holes*. The holes can recapture electrons from neighboring neutral atoms; therefore *hole migration* occurs in the materials system as a result of jumps of bound electrons between neighboring atoms. These bound electrons occupy energy states in the *valence band* of the semiconductor. Instead of analyzing the jumps of bound electrons, it is convenient to consider movement of holes as positive quasiparticles of charge $+e$. Like unbound electrons in the conduction band, the holes in the valence band can move freely and contribute to electrical conductance. Note that the energy states of freely moving electrons in the conduction band and holes in the valence band are separated by the 'forbidden' energy gap E_g (Fig. 5.11a). This energy gap is an essential factor in operation of optoelectronic devices.

If two regions with different types of conductance, i.e. n-type and p-type, are brought in contact, they form a *pn-junction* and the corresponding two-terminal device based on a *pn*-junction is called a *diode*. The *pn*-junction is characterized by energy barriers E_b (both for electrons and for holes) and

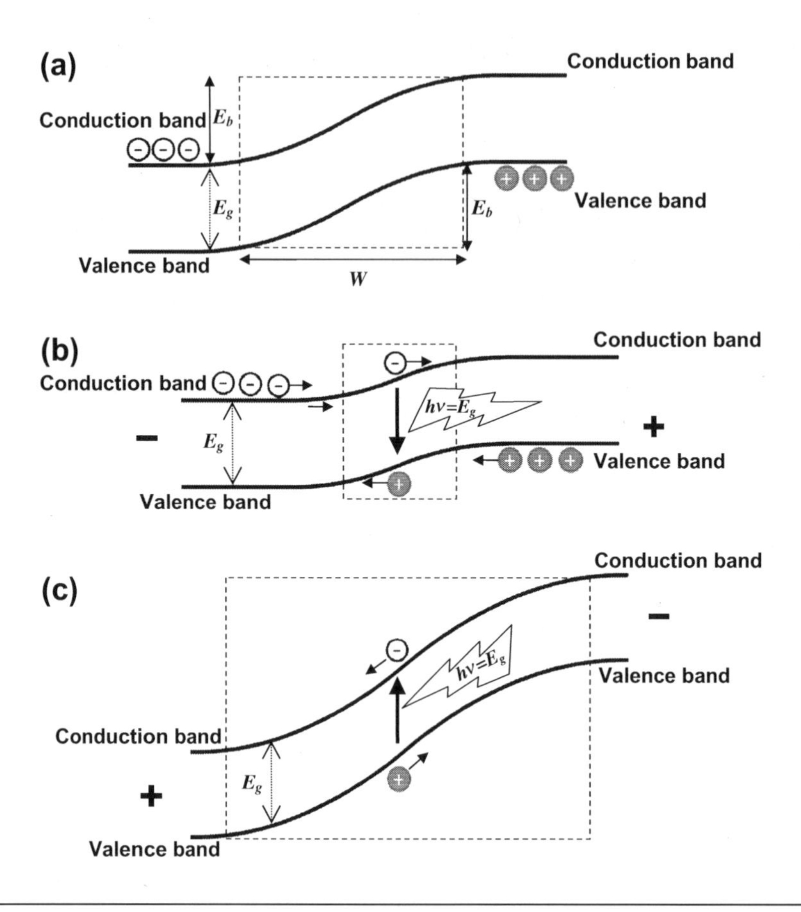

FIGURE 5.11

Typical energy band diagrams of LED and PD based on a pn-junction: (a) unbiased pn-junction; (b) forward biased LED; (c) reverse biased PD

by barrier length W (also called *depletion length* because this region is depleted of free carriers), as shown in Figure 5.11a. The barrier E_b prevents movement of electrons from the *n*-region into *p*-region and vice versa, movement of holes from the *p*-region into the *n*-region. The barrier (depleted) region of length W acts as the active region of the optoelectronic devices, where photons are generated (LED) or absorbed by producing an electrical signal (PD).

If now an electrical bias is applied to the *pn*-junction with the negative side on the *n*-region and positive side on the *p*-region, the external electric field opposes the direction of the built-in electric field of the *pn*-barrier. As a result, the barrier height is reduced and more electrons and holes can pass through the barrier region (Fig. 5.11b). In this *forward bias* regime, light can be generated in certain types of semiconductors. In the barrier region, an electron and hole can collide and the hole captures the electron forcing it to 'fall' from the conductance into the valence band, thus reducing its energy by E_g. Energy conservation requires that this energy must be released in some form. In light-emitting diodes, this occurs by emission of a photon with energy $h\nu \sim E_g$ (Fig. 5.11b).

A LED emits light only when the external bias voltage is above a certain 'threshold' level, V_{th}, also called turn-on voltage. The approximate relation between the photon wavelength and the turn-on voltage is

$$eV_{th} \sim E_g = \frac{hc}{\lambda} \tag{5.33}$$

The wavelength or 'color' of light is determined by the semiconductor bandgap E_g, which depends on the material's composition. Several examples of LED implementations are shown in Table 5.2. Typical materials used in LED emitting in the near-infrared range, i.e. $\lambda \sim 1$ µm, are GaAs and AlGaAs.

The *pn*-junction can be also used for detecting photons (the corresponding device is called a photodiode). This process is just the reverse of the light-generation process described above. When a photon is absorbed by a neutral atom in the barrier region, it can excite a bound electron to the conduction band, provided the energy of the absorbed photon $h\nu \geq E_g$. As a result an electron–hole pair is formed. The built-in electric field in the barrier region separates electrons and holes preventing them from re-annihilation. These generated electrons and holes are collected in the *n*- and *p*-regions respectively and a photocurrent is produced. The photodiode can operate in two modes. The first mode is when zero bias is applied to the *pn*-junction (Fig. 5.11a). This is called photovoltaic mode and is used, e.g., in solar cells. If instead an electrical bias is applied to the *pn*-junction with the positive side on the *n*-region and negative side on the *p*-region, the external electric field has the same direction as the direction of the built-in electric field of the *pn*-barrier. As a result, both the barrier height, E_b, and the length of the barrier region, *W*, are increased (Fig. 5.11c). This regime of operation is called *reverse bias*. The increased barrier region (active region) can absorb more photons (i.e. larger signal) and also exhibit a smaller junction capacitance resulting in faster response times. At higher reverse biases the photodiode can operate in *avalanche* regime when photogenerated carriers gain sufficient kinetic energy to excite additional electron–hole pairs, resulting in internal gain inside the photodiode. The avalanche PD can, in principle, detect even single photons [23].

Typical materials used in photodiodes are Si ($\lambda = 0.2$–1.1 µm), Ge ($\lambda = 0.4$–1.7 µm), and InGaAs ($\lambda = 0.8$–2.6 µm). The minimum detectable wavelength of EM radiation is determined by the bandgap of the semiconductor material.

Generic structures of a light-emitting diode and a photodiode are shown in Figure 5.12. Note that, in principle, semiconductor emitters and detectors of EM radiation are equivalent to two-dimensional arrays of many 'atomic antennas' [24]. Scaling limits of both LED and PD can be assessed by analyzing dimensions of the *optical window*, *d* (which is close to the device lateral dimensions), and the *active region*, *W*; this analysis follows.

Table 5.2 Examples of LED implementations

Color	Wavelength	LED turn-on voltage	Material
Near Infrared	$\lambda > 0.8$ µm	$V_f \sim 1.5$ V	GaAs, AlGaAs
Red	$\lambda = 0.61$–0.76 µm	$V_f \sim 2$ V	AlGaAs, GaAsP, AlGaInP
Green	$\lambda = 0.50$–0.57 µm	$V_f \sim 2.5$ V	InGaN, GaP
Blue	$\lambda = 0.45$–0.50 µm	$V_f \sim 3$ V	InGaN

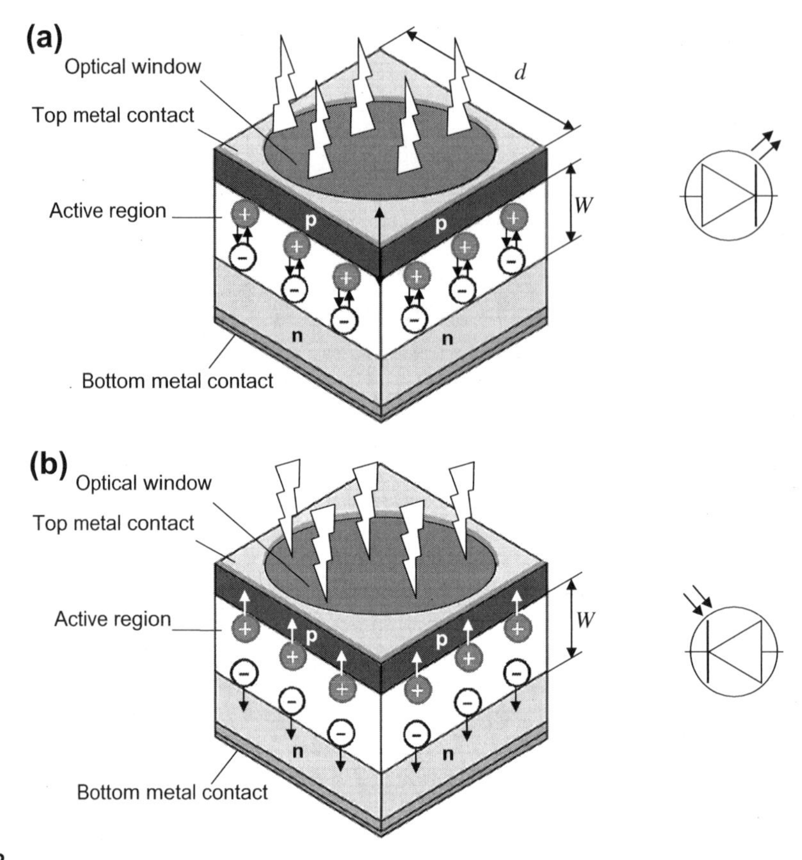

FIGURE 5.12

Generic structures of a light-emitting diode (a) and a photodiode (b)

5.9.2 Scaling limits of optoelectronic devices

Operation of all light-emitting and light-absorbing devices can be viewed in the context of a light beam passing through an aperture (optical window in Fig. 5.12). Scaling limits of these devices can be assessed based on the ability of photons to pass through the small optical windows. It can be intuitively argued that transmission through a small opening should be difficult if the opening diameter is less than the effective 'size' of photon, which is on the order of wavelength λ. Indeed, Bethe [25] calculated the transmission of light through a hole with radius $r << \lambda$ in an infinitely thin, ideally conducting film and found that the transmission falls proportionally to $(r/\lambda)^4$. Specifically, the expression for the Bethe transmission efficiency η_B (normalized to the aperture area) is [26]

$$\eta_B = \frac{64}{27\pi^2}(kr)^4 \tag{5.34}$$

Since $k = 2\pi/\lambda$, (5.34) can be re-written as

$$\eta_B = \frac{64}{27\pi^2}\left(\frac{2\pi}{\lambda}r\right)^4 = \frac{64}{27\pi^2}16\pi^4\left(\frac{r}{\lambda}\right)^4 \approx 374\left(\frac{r}{\lambda}\right)^4 \tag{5.35}$$

Note that the numerical factor in (5.35) is very close to $Z_0 = 377\ \Omega$, the impedance of free space, and the equation for the light transmission efficiency resembles the antenna equations discussed earlier in Section 5.4 (except that r/λ is now raised to the power of four rather than the power of two).

It is also interesting to note that application of (5.35) to a 'half-wave' size opening, with the hole diameter $d = 2r = \lambda/2$ results in $\eta_B = 1.46$, i.e. reasonably close to 1. This suggests that the functional form of (5.35) can be used as a first-order approximation for the transmission cut-off for $\lambda \geq 4r$ [26].

The transmission efficiency is further decreased, if the non-zero thickness of the opening is taken into account. Equations (5.34) and (5.35) were obtained with the assumption of zero-thickness, which holds when the hole thickness $\delta << \lambda$. For a simple estimate again apply the 'half-wave' condition, i.e. the opening thickness affects the transmission if $\delta \geq \lambda/2$. A 'thick' aperture can be represented as a series of 'thin' apertures, each of which is $\lambda/2$ thick. The transmission through this concatenation of apertures can be treated as a series of transmissions through $n = 2\delta/\lambda$ thin apertures, each having a transmission coefficient of (5.35). The total transmission coefficient in this case is

$$\eta(\delta) = \eta^n \propto \left[\left(\frac{r}{\lambda}\right)^4\right]^{\frac{2\delta}{\lambda}} \propto \left(\frac{r}{\lambda}\right)^{\frac{8\delta}{\lambda}} \tag{5.36}$$

and this results in a very sharp cut-off of light transmission through thick apertures.

It can be concluded that for energy-efficient operation, the spatial dimensions of optoelectronic devices should be larger than the incident wavelength. In fact the optical window should be larger than λ to minimize the effects of the aperture thickness (5.36). In the following, the minimum size of transmitting LED in a nanomorphic cell will be assumed to be $d = 2$ μm. The thickness of the active layer will be assumed to be $W \sim 1$ μm so this nanomorphic LED could be fit into a 2 μm cube.

An estimate of the energy needed to activate a LED can be made based on a simple LED circuit model shown in Figure 5.13. A control switch is used to connect a voltage source to the LED; thus pulses of light can be generated to send information. Indeed such a simple scheme could be suitable for the nanomorphic cell application due to the minimum number of components. Note that there is always an LED internal resistance and capacitance as shown in Figure 5.13. The resistance R is due to finite (usually relatively large) resistivity of semiconductor materials resulting in, e.g., noticeable contact

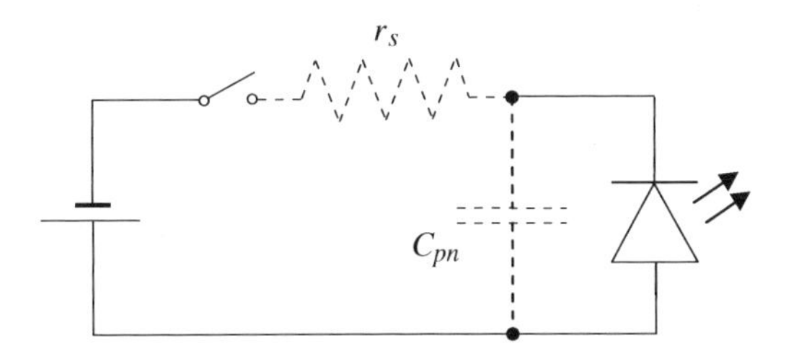

FIGURE 5.13

A simple LED circuit model

resistance. Also, all *pn*-junctions have *junction capacitance*, due to the finite size of the barrier length (depletion length) *W*, which acts as an effective insulating layer separating two 'plates' formed by *n*- and *p*-regions. The junction capacitance of a nanomorphic LED can be estimated as

$$C_{pn} \sim \frac{\varepsilon_0 K d^2}{W}. \tag{5.37}$$

A numerical estimate for a GaAs LED ($K_{GaAs} = 13.1$) with dimensions $d = 2$ μm and $W = 1$ μm gives $C_{pn} \sim 5 \times 10^{-16}$ F.

When the switch in Figure 5.13 is in the ON state, the LED is activated after the junction capacitor C_{pn} is charged to a voltage larger than the turn-on voltage V_{th} (see (5.33) and Table 5.2). Let $V_{th} \sim 1$ V; then the energy of charging the junction capacitor needed to activate the LED will be

$$E_{LED} = \frac{C_{pn}^2 V_{th}^2}{2} \sim \frac{5\text{x}10^{-16} \cdot 1^2}{2} = 2.5 \cdot 10^{-16} J \tag{5.38}$$

Equation (5.38) can be used as an optimistic estimate of the energy needed to send a bit of information using directional optical communication at $\lambda \sim 1$ μm.

It should be noted that in practical LED there are additional factors which severely limit efficiency. In fact a considerable portion of light generated in a LED is trapped inside by internal reflection due to a high refractive index of light-emitting semiconductor materials. This internal reflection limits the photon extraction efficiency. For example, a conventional planar InGaN light-emitting diode has the photon extraction efficiency of less than 5% [27]. Different approaches to enhance light extraction include the use of shaped die, rough surfaces or textured semiconductor surfaces, micro-lens arrays, and photonic crystals [27]. However all these techniques require an increase in the total size of the structure compared to the active region, and thus are difficult to implement in the nanomorphic cell.

5.10 STATUS OF μ-SCALED LEDs AND PDs

Currently, there is considerable interest in development of LEDs and PDs of several micrometers in size (μLED and μPD). Example application drivers for μLED are microdisplays, as well as different bioelectronic applications, such as lab-on-chip systems, neural stimulation, etc. [27–29]. There remain open questions regarding the performance of optoelectronic devices with device size scaling [28]. A recent example of the small 'conventional' μLEDs are arrays of InGaN disk-shaped light-emitting diodes with a diameter of 12 μm and emission wavelength of $\lambda = 408$ nm, i.e. $d/\lambda \approx 30$ [27].

A new class of light emitters, the light-emitting transistor, has been recently proposed [30,31] targeting high-speed optical data communication. Scaling properties of this light-emitting transistor have been investigated and devices with emitter aperture size as small as 5 μm have been demonstrated [31]. The peak emission wavelength of the device was ~ 1 μm, i.e. $d/\lambda = 5$.

'Unconventional' subwavelength LEDs based on quantum dots have also been reported, which target special applications, such as single-photon sources [32]. In [32], quantum dot (InAs/GaAs) light-emitting diodes were reported with an active area of 0.6 μm, emission wavelength of 1.3 μm ($d/\lambda \approx 0.5$), and efficiency of $\sim 0.01\%$ [32].

Developments in microphotodiodes are supported by several emerging bioelectronic applications, such as retinal implants for artificial vision [32,33], 'video pills' for visual inspection inside the human body [33], biomolecular recognition [35], etc. Recently an image sensor has been demonstrated [33], realized on 1.1×1.3 mm^2 silicon, which has a pixel array of 208×186 pixels. The pixel pitch was 4.6 μm and the size of optical window was about 2 μm. The maximum spectral sensitivity of these μPD was about 640 nm, thus $d/\lambda \approx 3$.

5.11 CONCLUDING REMARKS

This chapter reviewed the limits of *ubiquitous* communication for a nanomorphic cell. Electromagnetic radiation is the primary physical effect used for such ubiquitous communication and this chapter has focused on the basic physics of information transmission via electromagnetic radiation. Estimates were offered for the size/volume and energy required for communication in the nanomorphic cell environment. Discussion in the chapter was devoted primarily to transducers of electrical excitation into electromagnetic waves and vice versa. While there are a number of different components in an EM communication system, it is argued that the size and efficiency of a transducer are the gating factors for nanomorphic cell communications. Therefore, the limiting analyses offered in this chapter center on the electromagnetic transducer (e.g. antenna, LED, etc.) since its scaling directly impacts communication system performance.

Based on the physics of electromagnetic radiation, the primary physical parameter, which determines the scaling limits of the communication system, is the *radiation wavelength* λ compared with the size of the nanomorphic cell, L_{cell}. A condition for an efficient EM transmission is $\lambda \sim L_{cell}$, and therefore the size of the nanomorphic cell pre-determines the choice of the radiation wavelength used for communication.

If omnidirectional (uniform) radiation from the antenna is required, then an argument is presented for the energy required to uniformly 'paint' a sphere with information-carrying photons whose size is defined by the transmission wavelength. It was shown that the energy required to provide one photon uniformly over the surface of the receiving sphere depends inversely on the cube of the transmitted wavelength. When the transmitted wavelength is commensurate with the size of the nanomorphic cell (about ten microns), then it follows that the energy required for omnidirectional radiation of distance one meter is about nine orders of magnitude greater than the energy required to operate one logic switch.

Larger EM radiation wavelength would, in principle, reduce the energy of omnidirectional communication. However, if the communication wavelength is much larger than the antenna length, the radiation efficiency of the antenna is very low – a crucial concern for the energy-limited nanomorphic cell.

Another difficult problem for nanomorphic communication is the *background thermal radiation*. Interestingly, at room temperature, the thermal radiation spectrum dramatically peaks around the ten-micron wavelength. This means that electromagnetic radiation for wavelengths in proximity to ten microns must carry sufficient energy, i.e. enough photons, to overcome the background thermal noise. This is again a challenge for the energy-constrained nanomorphic cell. Therefore it is necessary to explore communication options beyond the thermal radiation region (circa 5–50 nm). One option is to use larger wavelength, e.g. $\lambda > 100$ μm (THz range), and the other option is to go a much shorter wavelength, e.g. $\lambda \sim 1$ μm (near infrared optical range).

In the case of THz radiation, with wavelengths greater than one hundred microns (an external antenna of approximately the same size is required), the energy for omnidirectional communication could be substantially lower for the same transmission distance than for ten-micron wavelength transmission. Unfortunately, if the nanomorphic cell is used in an in vivo application, one encounters a limitation due to the strong absorbance of water in this range of wavelengths. Another grand challenge for the utilization of THz radiation is that the integrated circuit technology has not yet been developed to support signal generation and detection.

The other possibility that avoids the peak of the thermal noise spectrum is the use of optical wavelengths, e.g., on the order of one micron (near-infrared). Existing optoelectronic technologies could in principle support operation at ~ 1 μm wavelength. However, omnidirectional communication by the nanomorphic cell is practically forbidden for this regime, since the energy required to send a bit of information at $\lambda = 1$ μm is close to the total energy budget of the nanomorphic cell. Different schemes of directional transmission could reduce the number of photons in the packet and therefore the total energy consumed. However, the orientation problem must be solved, which might require additional energy expenditure. Alternatively, directional transmission could be augmented by using as many external detectors as possible or even a surround of receivers (e.g. MRI-like). This certainly limits the application space but a general conclusion is that the optical wavelengths do not allow *ubiquitous communication by the nanomorphic cell*. However, in contrast with the THz regime, generation and detection of optical (~ 1 μm) radiation allows for the use of devices whose physical size is commensurate with the dimensions of the nanomorphic cell. A brief review of the physics of the operation of optoelectronic devices, such as light-emitting diodes and photodiodes, was given in this chapter followed by an analysis of the scaling limits for optoelectronic devices. While there are many factors limiting the efficiency of optoelectronic devices, the scaling limits of these devices universally depend on the ability of light to pass through an aperture/optical window under which they are placed. It was argued that the size of the optical window should be larger than the radiation wavelength for an efficient light transmission. The estimates of the minimum energy for the 1-μm directional communication are 6–7 orders of magnitude less than for the omnidirectional communication schemes. However, the energy to send a bit is still about two orders of magnitude larger than the energy required to operate one logic switch. Thus, comparisons of the energy costs to 'process' one bit with the communication energy costs to transmit one bit of information suggests that the overall design goal should be to minimize communication and to maximize the 'intelligence' of the nanomorphic cell.

LIST OF SYMBOLS

Symbol	Meaning
c	Velocity of light in vacuum, $c = 3 \times 10^8$ m/s
C	Capacitance
C_{pn}	*pn*-junction capacitance
C_{wire}	Wire capacitance
d	Diameter
e	Electron charge, $e = 1.6 \times 10^{-19}$ C
E	Energy

Symbol	Meaning
E_b	Energy barrier height
E_{com}	Communication energy per bit
E_g	Semiconductor bandgap
E_{LED}	LED turn-on energy
E_{ph}	Photon energy
E_{stored}	Stored energy
E_{SW}	Switching energy
h	Planck's constant, $h = 6.63 \times 10^{-34}$ J·s
I, I_0	Current
k	Wave number, $k = 2\pi/\lambda$
k_B	Boltzmann constant, $k_B = 1.38 \times 10^{-23}$ J/K
K	Dielectric constant
L	Distance, length
L_{ant}	Length of antenna
L_{cell}	Length of nanomorphic cell
L_{opt}	Optimum antenna length
n	Integer
$N_{4\pi}$	Number of photons emitted into solid angle of 4π steradians
N_{bit}	Number of bits
N_{ph}	Number of photons
p, p_0	Radiation power density (intensity)
P_{in}	Input power
P_{loss}	Loss power
P_n	Noise power
P_{ph}	Detected photon power
P_{rad}	Radiation power
P_s	Signal power
r	Radius
R	Resistance
R_{loss}	Loss resistance
R_{rad}	Radiation resistance
$R_{\lambda/2}$	Radiation resistance of half-wave antenna, $R_{\lambda/2} = 73.1\ \Omega$
SNR	Signal-to-noise ratio
t, τ	Time, time interval
t_H	Heisenberg time
t_s	Signal time
T	Absolute temperature
u	Velocity
V, $V(t)$, V_Δ, V_0	Voltage
V_{th}	Threshold/turn-on voltage
W	Barrier length/depletion length

(Continued)

Symbol	Meaning
Z_0	Free space impedance, $Z_0 = 377\ \Omega$
α	Absorption coefficient
$\Delta\nu$	Bandwidth
δ	Thickness
ε_0	Permittivity of free space, $\varepsilon_0 = 8.85 \cdot 10^{-12}$ F/m
η	Transmission coefficient
η_B	Bethe transmission efficiency
η_{rad}	Radiation efficiency
Θ	Period of oscillation
λ	Wavelength
ν	Frequency
τ	Time interval
$\propto$	Indicates proportionality
$\sim$	Indicates order of magnitude

References

[1] D. Gabor, Communication theory and physics, Phil. Mag., ser.7 41 (1950) 1161–1187.
[2] L.V. Blake, Antennas, ARTECH HOUSE, Inc, 1984.
[3] J.D. Jackson, Classical Electrodynamics, John Wiley & Sons, Inc, 1998.
[4] G.W. Hanson, Radiation efficiency of nano-radius dipole antennas in the microwave and far-infrared regimes, IEEE Antenn. Propag. Mag. 50 (2008) 66–77.
[5] A. Karlsson, Physical limitations of antennas in a lossy medium, IEEE Trans. Antenn. Propag. 52 (2004). 2037–2033.
[6] H.T. Friis, A note on a simple transmission formula, Proc. IRE 34 (1946) 254–256.
[7] B.W. Cook, S. Lanzisera, K.S.J. Pister, SoC issues for RF smart dust, Proc. IEEE 94 (2006) 1177–1196.
[8] J. Ammer, J. Rabaey, The energy-per-useful-bit metric for evaluating and optimizing sensor network physical layers, Proc. IEEE Intern. Workshop on Wireless Ad Hoc & Sensor Networks (2006) 695–700.
[9] Q. Zhu, L. Wu, S. Sheng, Z.C. Mei, W.F. Liu, W.L. Cai, et al., Possibility of constructing microwave antenna with carbon nanotubes, J. Vac. Sci. Techol. B 25 (2007) 1630–1634.
[10] E.H. Putley, The detection of sub-mm radiation, Proc. IEEE 51 (1963) 1412–1423.
[11] THz-BRIDGE: Tera-Hertz radiation in Biological Research, Investigation on Diagnostics and study of potential Genotoxic Effects, http://www.frascati.enea.it/THz-BRIDGE/
[12] D.C. Yates, A.S. Holmes, Preferred transmission frequency for size-constyrained ultralow-power short-range CMOS oscillator transmitters, IEEE Trans. Circ. Syst. I 56 (2009) 1173–1181.
[13] S. Soora, K. Gosalia, M.S. Humayun, G. Lazzi, A comparison of two and three dimensional dipole antennas for an implantable retinal prosthesis, IEEE Trans. Antenn. Propag. 56 (2008) 622–629.
[14] J. Abadia, F. Merli, J.-F. Zurcher, J.R. Mosig, A.K. Skrivervik, 3D-spiral small antenna design and realization for biomedical telemetry in the MICS band, Radioengineering 18 (2009) 359–367.
[15] S. Sufyar, C. Delaveaud, A miniaturization technique of a compact omnidirectional antenna, Radioengineering 18 (2009) 373–379.

[16] J. Thoné, S. Radiom, D. Turgis, R. Carta, G. Gielen, R. Puers, Design of a 2 Mbps FSK near-field transmitter for wireless capsule endoscopy, Sensors and Actuators A 156 (2009) 43–48.

[17] N. Boom, W. Rens, J. Crols, A 5.0 mW 0 dBm FSK transmitter for 315/433 MHz ISM applications in 0.25 mm CMOS, ESSCIRC 2004: Proc. 30th European Solid-State Circ. Conf. (Sept. 21–23, 2004, Leuven, Belgium) 199–202.

[18] M.W. Shen, C.Y. Lee, J.C. Bor, A 4.0-mW 2-mbps programmable BFSK transmitter for capsule endoscopy applications, Proc. 2005 IEEE Asian Solid-State Circ. Conf. (Nov. 1–3, Hsinchu. Taiwan) 245–248.

[19] B.Y. Chi, J.K. Yao, S.G. Han, X. Xie, G.L. Li, Z. Wang, Low-power transceiver analog front-end circuits for bidirectional high data rate wireless telemetry in medical endoscopy applications, IEEE Trans. Biomed. Eng. 54 (2007) 1291–1299.

[20] Y.H. Liu, T.H. Lin, An energy-efficient 1.5 Mbps wireless FSK transmitter with a Sigma Delta-modulated phase rotator, ESSCIRC 2007: Proc. 33d European Solid-State Circ. Conf. (Sept. 11–13, 2007, Munich, Germany) 488–491.

[21] T.M. Hancock, M. Straayer, A. Messier, A sub-10 mW 2 Mbps BFSK transceiver at .35 to 1.75GHz, Proc. IEEE Radio Freq. Integr. Circ. (RFIC) Symp. (June 3–5, 2007, Honolulu, HI) 97–100.

[22] S.M. Sze, Physics of Semiconductor Devices, John Wiley and Sons, 1981.

[23] F. Zappa, A.L. Lacaita, S.D. Cova, P. Lovati, Solid-state single-photon detectors, Optical Eng. 35 (1996) 938–945.

[24] J.M. Kahn, J.R. Barry, Wireless infrared communications, Proc. IEEE 85 (1997) 265–298.

[25] H.A. Bethe, Theory of diffraction by small holes, Phys. Rev. 66 (1944) 163–182.

[26] C. Genet, T.W. Ebbesen, Light in tiny holes, Nature 445 (2007) 39–46.

[27] Z.Y. Fan, J.Y. Lin, H.X. Jiang, III-nitride micro-emitter arrays: development and applications, J. Phys. D. : Appl. Phys. 41 (2008) 094001.

[28] Z. Gong, S. Jin, Y. Chen, J. McKendry, D. Massoubre, I.M. Watson, E. Gu, M.D. Dawson, Size-dependent light output, spectral shift, and self-heating of 400 nm InGaN light-emitting diodes, J. Appl. Phys. 107 (2010) 013103.

[29] V. Poher, N. Grossman, G.T. Kennedy, K. Nicolic, H.X. Zhang, Z. Gong, et al., Micro-LED arrays: a tool for two-dimensional neuron stimulation, J. Phys. D: Appl. Phys. 41 (2008).

[30] M. Feng, N. Holonyak, W. Hafez, Light-emitting transistor: Light emission from InGaP/GaAs hetero-junction bipolar transistors, Appl. Phys. Lett. 84 92004) 151–153.

[31] C.H. Wu, G. Walter, H.W. Then, M. Feng, N. Holonyak, Scaling of light emitting transistor for multi-gigahertz optical bandwidth, Appl. Phys. Lett. 94 (2009) 171101.

[32] C. Monat, B. Alloing, C. Zinoni, L.H. Li, A. Fiore, Nanostructured current-confined single quantum dot light-emitting diode at 1300 nm, Nano Lett. 6 (2006) 1464–1467.

[33] E. Zrenner, Will retinal implants restore vision? Science 295 (2002) 1022–1025.

[34] G. H-Graf, C. Harendt, T. Engelhardt, C. Scherjon, K. Warkentin, H. Richter, J.N. Burghartz, High dynamic range CMOS imager technologies for biomedical applications, IEEE J. Solid-State Circ. 44 (2009) 281–289.

[35] J.P. Conde, A.C. Pimentel, A.T. Pereira, A. Gouvêa, D.M.F. Prazeres, V. Chu, Detection of molecular tags with an integrated amorphous silicon photodetector for biological applications, J. Non-Cryst. Solids 354 (2008) 2594–2597.

CHAPTER

Micron-sized systems: In carbo vs. in silico

6

CHAPTER OUTLINE

6.1 INTRODUCTION

In Chapters 2–5, scaling properties of essential units of autonomous electronic systems have been investigated: energy sources, information processing (logic and memory), sensing, and communication. The reader is now equipped with the quantitative estimates of performance parameters for each of the above units, when the size of the unit is ~10 μm. The next task is to assemble all units into an integrated system and to analyze the performance of this in silico system, assuming a best case scenario.

It is intriguing to compare the ultimately scaled in silico systems with natural in carbo systems, i.e. living cells. In fact, the living cell is an excellent example of a functioning micron-sized information-processing system and it will be used as a benchmark for comparison with operation of a nanomorphic cell, a micron-sized electronic system. Therefore, a goal of this chapter is to develop a framework for estimating the information-processing capabilities both of a living cell and a nanomorphic cell.

As prerequisites for quantitative assessments of the in silico and the in carbo functional microsystems, theoretical models for information and information processors are introduced in this chapter. Then the living cell is characterized as an information-processing system in the context of the theoretical framework of a *Turing Machine* and a *von Neumann Universal Constructor.*

The study of the living cell as a functional microsystem may help engineers understand the limits of scaling for functional electronic systems and offer new insights that avoid some of the restrictions of classical approaches to information-processing devices and architectures. Conversely, lessons from the extremely scaled electronic systems may help biologists gain new insights on how cells function.

BOX 6.1 IN SILICO AND IN CARBO SYSTEMS

In silico, in its original meaning, is a phrase used to mean 'performed on computer' in analogy to in vivo, which is commonly used in biology and refers to processes occurring within living organisms. In this text, the expression in silico is used in a broader sense, referring to different processes occurring in an electronic microsystem (which is to a large extent made from silicon). This is in contrast to similar processes in living cells, which are to a large extent made of carbon, thus being in carbo systems.

6.2 INFORMATION: A QUANTITATIVE TREATMENT

The concept of information has been discussed qualitatively in Chapter 3, with respect to different information carriers. In this section, a quantitative definition of information is introduced. The simplest special case of equal probability of occurrence of information events is initially considered to provide an introduction to mathematical information concepts. For a general and rigorous mathematical treatment of the concept of information, the reader is referred to treatises on information theory, for example [1–5].

6.2.1 An intuitive introduction to information theory

This section follows the introductory approach to information theory used by Leon Brillouin [2]. Assume that N different events can occur and that all N events have equal probability of occurrence. Physically occurring information events can be represented as *symbols* encoded in states of physical matter (e.g. distinguishable states created by the *presence* or *absence* of material particles). The *information* function $I(N)$ then can be informally defined as a measure of the likelihood that one of N events will occur (e.g. a given symbol to appear). The less likely an event is, the larger is the surprise at its occurrence, and thus the information gained. For N events with equal probability of occurrence, the probability, p, of realization of one out of N events, and thus the likelihood of the event, is:

$$p = \frac{1}{N} \tag{6.1}$$

Obviously, the larger N is, the smaller is the probability of occurrence of a selected event, and therefore the larger is the information gained. Thus, information should be an increasing function of N. In contrast, if $N = 1$, only one event can occur (e.g. just one symbol, always present) – there is no uncertainty and therefore no information is gained from the observation of this event, i.e. $I(1) = 0$.

It is convenient to require that the information exhibit an *additive property*, i.e. if information is received from different independent sources, the total information is the sum of the information acquired from each source. For example, consider a situation where a decision is made based on several, say two, inputs and each input has several independent realizations; for example, Input 1 has K realizations and Input 2 has L realizations. In this case, the total number of possible outcomes, from the *Fundamental Principle of Counting*, is $N = K \cdot L$.

Fundamental Principle of Counting: If there are m ways to do one thing, and n ways to do another, then there are (m × n) ways of doing both.

The additive property of information requires that

$$I(K \cdot L) = I(K) + I(L) \tag{6.2}$$

We thus seek a function that satisfies the equation:

$$f(x_1 \cdot x_2) = f(x_1) + f(x_2) \tag{6.3a}$$

the solution of which is

$$f(x) = \log x \tag{6.3b}$$

(It can be proven rigorously, that logarithm is the only mathematical function satisfying (6.3a) – for the proof see [3].) Now, an *information function* satisfying the above requirements is

$$I(N) = C \log N \tag{6.4}$$

where C is a constant. The choice of the base of the logarithm and the constant C depends on the system of units. The natural logarithm (ln), decimal logarithm (lg), and logarithm to the base-2 (ld) are the most convenient choices.

6.2.2 Units of information

For a quantitative treatment of information, a system of units of $I(N)$ in (6.4) is needed. Note that the log base and the pre-log constant are interdependent: two logarithms with different bases b_1 and b_2 for the same argument are directly proportional to each other:

$$\begin{aligned} &\log_{b_1} x \propto \log_{b_2} x \\ &\log_{b_1} x = \log_{b_1} b_2 \cdot \log_{b_2} x = C' \cdot \log_{b_2} x \end{aligned} \tag{6.5}$$

where $C' = \log_{b_1} b_2$.

In principle, different logarithm bases can be used, depending on the application. Most universal units are associated with the minimum number of alternatives sufficient for a decision process. The smallest number of *different* symbols $N_{min} = 2$ (the binary choice). It is convenient to require that the informational content of the binary choice equals to:

$$I(2) = 1 \tag{6.6a}$$

Correspondingly,

$$C \log_b 2 = 1 \tag{6.6b}$$

or

$$C = \frac{1}{\log_b 2} \tag{6.6c}$$

Now, to eliminate the pre-log constant, require $C = 1$ and, from (6.6c), obtain the base $b = 2$. Now the final expression for binary information is

$$I(N) = \log_2 N \tag{6.7}$$

The unit of information defined by (6.7) is called a *bit* (binary digit).

Example 1: Information content of a binary word of length n (i.e. consisting of n '1's or '0's). The total number of possible combinations in this case is $N = 2^n$, and therefore:

$$I(n) = \log_2 2^n = n \log_2 2 = n \text{ bit} \tag{6.8}$$

Example 2: Information content of English alphabet (upper bound).

The English alphabet contains 26 letters. Consider written text as a sequence of events (each letter is one event): then each event will have 27 possible outcomes (including the blank space).

Assuming equal probability of occurrence for each letter, the information per symbol for $K = 27$ using (6.7) is:

$$I = \log_2 27 = 4.75 \text{ bit/letter}$$

In actual use, different letters have an unequal probability of occurrence (see Example 3 in Section 6.2.3 below). As will be shown in Section 6.2.3, the number obtained in Example 2, represents the upper bound for information per symbol.

Example 3: Information content of proteins (lower bound) [5].

All living organisms use the same 20 amino acids (called common or standard amino acids) [6] as building blocks for the assembly of protein molecules (see Table 6.1). They form the 'alphabet' of proteins, the main building blocks of living systems. Proteins are made from long (e.g. hundreds of units) sequences of different amino acid 'symbols', which are folded in complex 3D arrangements. For example, the average size of proteins in *E. coli* bacteria is 360 'symbols' [9] and the structural information of the corresponding linear chain of amino acids is

$$I_A = \log_2 20^{360} = 360\log_2 20 \approx 1556 \text{ bit}$$

It is important to note that the above number represents the lower bound and the information content of the 3D protein structure is much greater than that of the amino acid sequence I_A [5].

Note that in the Examples 2 and 3, the information content of a non-binary system (i.e. $N = 27$ and $N = 20$) was presented with binary units. This applies to an arbitrary non-binary system – the information content of a system with an arbitrary number of outcomes can be described using binary

Table 6.1 Standard amino acids: Symbols of the 'alphabet' of Life

Name	Molecular formula	Abbreviation	3-letter abbreviation
Alanine	$C_3H_7NO_2$	A	Ala
Arginine	$C_6H_{14}N_4O_2$	R	Arg
Asparagine	$C_4H_8N_2O_3$	N	Asn
Aspartic acid	$C_4H_7NO_4$	D	Asp
Cysteine	$C_3H_7NO_2S$	C	Cys
Glutamic acid	$C_5H_9NO_4$	E	Glu
Glutamine	$C_5H_{10}N_2O_3$	Q	Gln
Glycine	$C_2H_5NO_2$	G	Gly
Histidine	$C_6H_9N_3O_2$	H	His
Isoleucine	$C_6H_{13}NO_2$	I	Ile
Leucine	$C_6H_{13}NO_2$	L	Leu
Lysine	$C_6H_{14}N_2O_2$	K	Lys
Methionine	$C_5H_{11}NO_2S$	M	Met
Phenylalanine	$C_9H_{11}NO_2$	F	Phe
Proline	$C_5H_9NO_2$	P	Pro
Serine	$C_3H_7NO_3$	S	Ser
Threonine	$C_4H_9NO_3$	T	Thr
Tryptophan	$C_{11}H_{12}N_2O_2$	W	Trp
Tyrosine	$C_9H_{11}NO_3$	Y	Tyr
Valine	$C_5H_{11}NO_2$	V	Val

units. Let the system have N outcomes, $N > 2$. We can present N as $N = 2^g$, where g is a real number. Now, using (6.4), obtain:

$$I(N) = C \log N = C \log 2^g = Cg \log 2 \quad (6.9)$$

and thus:

Main Point I:

Information of different types can be represented in the binary form, i.e. by bits!

Thermodynamic units

Another unit system appears from the similarity between (6.9) and the famous Boltzmann's formula for entropy used in thermodynamics (this equation is inscribed on Boltzmann's tomb):

$$S = k \log W = k_B \ln W \quad (6.10a)$$

where W is the number of possible realizations or microstates in a material system and k_B is Boltzmann's constant, one of the fundamental physical constants, $k_B = 1.38 \times 10^{-23}$ J/K.

In a system with only two microstates, $W = 2$, the information content of the system in thermodynamic units (J/K) is

$$I = k_B \ln 2 \quad (6.10b)$$

Note that the mathematical definitions of information (6.4) and thermodynamic entropy (6.10a) are identical in form. This has an important implication as it connects the concept of information to physics. For example, it allows for estimates of the information content of arbitrary material systems [4], including living organisms – an idea developed further in later sections of this chapter.

6.2.3 Optimum base for computation

As was shown in the previous section, information can be mathematically presented as a logarithm function of different bases (corresponding to a different number of symbols in the 'alphabet'). It is interesting to consider the most economical representation for a number, N. There are two factors contributing to the cost of numeric representation. The first factor is the number of symbols in the alphabet (or number of different numeric digits) b, which is the base of the system. The second factor is the length w of the sequence of the symbols needed, e.g. to represent a number in the range between 0 and N. For a given b and w the range is

$$N = b^w \quad (6.11)$$

Correspondingly, the length of the sequence needed to represent the range from 0 to N is

$$w = \log_b N \quad (6.12)$$

Both b and w need to be minimized in order to obtain an efficient representation for numbers with range N. To do this, let the *information efficiency* function Ψ be the product of b and w:

$$\Psi = bw \quad (6.13)$$

To find the minimum of (6.13), the derivative of (6.13) must be zero:

$$\Psi' = (bw)' = 0 \tag{6.14}$$

Substituting (6.12) into (6.14), there results:

$$(b \log_b N)' = 0 \tag{6.15}$$

Using the rule of changing base, obtain:

$$\log_b N = \frac{\ln N}{\ln b} \tag{6.16}$$

Now the derivative of (6.15) with respect to b, using (6.16), is:

$$(b \log_b N)' = \log_b N + b(\log_b N)' = \ln N\left(\frac{1}{\ln b} - b\frac{1}{b}\frac{1}{\ln^2 b}\right) = \ln N\left(\frac{\ln b - 1}{\ln^2 b}\right) \tag{6.17}$$

Equating (6.17) to zero, there results:

$$\ln b - 1 = 0 \tag{6.18a}$$

or

$$b_{opt} = e = 2.71828... \tag{6.18b}$$

which is the theoretical optimum base for computation. In practice, the base (the number of the symbols in the 'alphabet') must be an integer, thus the nearest integer base is $b_{opt} = 3$. The graph of the function $bw = b \log_b N$ is shown in Figure 6.1. As can be seen from the graph, for an economic representation of information, the computational base should be chosen among lower numbers, e.g. from 2 to 4. In addition to the 'economy principle' discussed above, there can be additional criteria for choosing the appropriate base of computation. For example, in electronic computing, base-2 is used (since 'ON' and 'OFF' states of an electronic switch can be used to represent binary numbers).

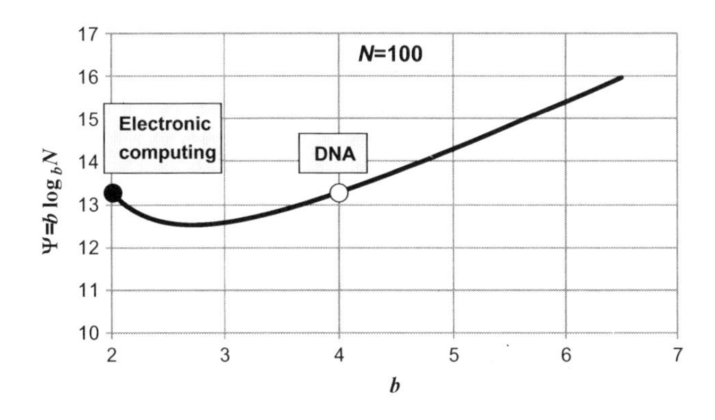

FIGURE 6.1

Information efficiency function $\Psi(b) = b\log_b N$ as a function of the computational base, b (the plot corresponds to $N = 100$). Note that $\Psi(2) = \Psi(4)$

On the other hand, in living organisms, base-4 is used to encode information in DNA, possibly for better error protection. As an interesting observation, as can be seen in Figure 6.1, the information efficiency functions are equal for the bases 2 and 4: $\Psi(2) = \Psi(4)$. This can be shown as follows:

$$\Psi(4) = 4 \log_4 N = 4 \cdot \frac{\log_2 N}{\log_2 4} = 4 \cdot \frac{\log_2 N}{2} = 2 \log_2 N = \Psi(2) \tag{6.19}$$

Main Point II:

The information efficiency of the computational base-2 used by in silico electronic computing is equal to the base-4 used by in carbo systems.

6.2.4 General case: Non-uniform probability of occurrence of information events

The concept of information also encompasses non-equal probabilities of occurrence for different symbols (events). Repeating the arguments in the beginning of this section, information is a measure of the likelihood that one of *N* events will occur (e.g. a given symbol to appear). The 'likelihood' in turn can be quantitatively described as the probability of occurrence of a given event, p_i, $i = 1,.., N$. Thus one can define the information function as

$$I(p_i) = -C \log p_i \tag{6.20}$$

The minus sign in (6.20) is due to the fact that the probability $p \leq 1$, and thus $\log p \leq 0$. It is useful for the information function to be non-negative, i.e. $I(p) \geq 0$, thus the minus sign. The information function (6.20) has the additive property, required earlier. Suppose a decision is to be made based on the joint occurrence of two independent inputs, characterized by the probabilities p_1 and p_2. The probability of a joint event 1&2 is the product $p_{12} = p_1 \times p_2$, and the corresponding information gain is

$$I(p_{12}) = I(p_1 \cdot p_2) = -C \log p_1 p_2 = -C \log p_1 - C \log p_2 = I(p_1) + I(p_2) \tag{6.21}$$

Note that the function *I(p)* (6.21) is consistent with the earlier obtained function *I(N)* (6.4) in the case of equal probability of occurrence of all *N* events ($p = 1/N$). Indeed:

$$I(p) = -C \log p = -C \log \frac{1}{N} = C \log N = I(N) \tag{6.22}$$

Next, for the arbitrary probabilities of occurrence of *N* symbols, the average information content per symbol can be calculated using standard formulae for weighted mean (or expected value):

$$\langle I \rangle = \sum_i^N p_i I_i(p_i) = \sum_i^N -C p_i \log p_i = -C \sum_i^N p_i \log p_i \tag{6.23}$$

This is a famous formula due to Claude Shannon, known as Shannon's entropy equation.

An important property of the average information function (6.23) is that it attains a maximum when all the probability values p_i are equal. For example, for $N = 2$, (6.23) becomes:

$$\begin{aligned} \langle I(p_1, p_2) \rangle &= -C(p_1 \ln p_1 + p_2 \ln p_2) \\ p_1 + p_2 &= 1 \end{aligned} \tag{6.24}$$

Table 6.2 The probability of occurrence of different letters in English

Symbol	Probability of occurrence, *p*	Symbol	Probability of occurrence, *p*	Symbol	Probability of occurrence, *p*
A	0.063	J	0.001	S	0.052
B	0.0175	K	0.003	T	0.047
C	0.023	L	0.029	U	0.0225
D	0.035	M	0.021	V	0.008
E	0.105	N	0.059	W	0.012
F	0.0225	O	0.0654	X	0.002
G	0.011	P	0.0175	Y	0.012
H	0.047	Q	0.001	Z	0.001
I	0.055	R	0.054		0.214

Let $p_1 = x$, and therefore $p_2 = 1 - x$:

$$\langle I(x)\rangle = -C(x \ln x + (1-x)\ln(1-x)) \tag{6.25}$$

Seeking $\frac{d\langle I(x)\rangle}{dx} = 0$, the maximum of $\langle I(x)\rangle$ is found as follows:

$$\begin{aligned} \frac{d\langle I\rangle}{dx} &= -C\frac{d}{dx}(x \ln x + (1-x)\ln(1-x)) = -C(\ln x - \ln(1-x)) = 0 \\ \ln x - \ln(1-x) &= 0 \\ x &= \tfrac{1}{2} \end{aligned} \tag{6.26}$$

Thus, the function $\langle I(p_1, p_2)\rangle = \langle I(p_1)\rangle$attains it maximum when $p_1 = p_2 = \frac{1}{2}$.

The proof for $N = 2$ can be generalized for an arbitrary number of symbols (see Appendix).

Example 4: Information content of English alphabet II (a detailed calculation).

In a real language, different letters have different probabilities of occurrence. The probability of occurrence of different letters in English is shown in Table 6.2. Applying Shannon's Equation (6.23) and using Table 6.2, obtain:

$$\langle I\rangle = -\sum_{i=1}^{27} p_i \log_2 p_i = 4.03 \text{ bit/letter}$$

Main Point III:

When the probability of occurrence of different symbols is not known, it is still possible to calculate the upper bound for the average information function.

Note that the above result (4.03 bit/letter) is reasonably close to the estimate in Example 2 where equal probabilities of occurrence was assumed (i.e. 4.75 bit/letter).

6.2.5 Information content of material systems

The overall information content of a material system consists of information about the system's composition and shape [4,7]. Consider a computer with the task of controlling the assembly of a structure from building blocks. Let the total number of the building blocks be K and assume there are N different kinds of building blocks. Next, assume a serial process with one building block added to the assembled structure per step. In the general case, for each step the computer must: (a) select the appropriate category of the building blocks, and (b) calculate x-, y-, z-coordinates of the position for each of the building blocks. According to (6.7), the information associated with the selection process (a) is $\log_2 N$, while information associated with 3-dimensional positioning is $3n$, where n is the length of the binary number representing each coordinate. A number length, $n = 32$ bits will be used in all subsequent treatments (it is usually sufficient for representing arbitrary numbers in a '*floating-point*' format). Now, the information per assembly step is

$$I_1 = \log_2 N + 3n \tag{6.27}$$

And the total information required for the assembly is

$$I_M = K(\log_2 N + 3n) \tag{6.28}$$

6.2.6 Information content of a living cell

Next imagine that individual atoms are to be assembled to construct a living cell. The *E. coli* bacterium will be used as a model cell because the most detailed information about its cellular composition is available (see [9–11]).

The elemental composition of a typical bacterial cell, such as *E. coli* [10], is shown in the second column of Table 6.3. The corresponding number of atoms can be calculated from the mass of the cell. The cell wet weight of *E. coli* is ~10–15 kg = 10–12 g. The cell dry weight is usually assumed to be

Table 6.3 Estimated composition of *E. coli* [10]

Element	% of dry weight	m, kg	N_{at}
C	50	2×10^{-16}	8×10^{9}
O	20	6×10^{-17}	2×10^{9}
N	14	4×10^{-17}	2×10^{9}
H	8	2×10^{-17}	1×10^{10}
P	3	9×10^{-18}	2×10^{8}
S	1	3×10^{-18}	6×10^{7}
K	1	3×10^{-18}	5×10^{7}
Mg	0.5	2×10^{-18}	4×10^{7}
Ca	0.5	2×10^{-18}	2×10^{7}
Fe	0.2	6×10^{-19}	6×10^{6}
Total	98.2%	3×10^{-16}	3×10^{10}

30% of the wet weight and thus is ~3 × 10–13 g. Multiplying this number by the percentage of a given element in the composition yields the total mass, m_{at}, of corresponding atoms in the cell (third column in Table 6.3). The number of atoms of the given element can be calculated based on:

$$N_{at} = \frac{m_{at}}{(1\ amu) \cdot M} \tag{6.29}$$

Where 1 $amu = 1.66 \times 10^{-24}$ g is the atomic mass unit and M is the molar mass of the corresponding element (see Box 2.2 in Chapter 2). The resulting number of atoms in the cell is given in the fourth column of Table 6.3.

This estimate results in the total number of atoms in an *E. coli* cell ~ 3×10^{10}.

Next, the the amount of information that needs to be processed to assemble a new cell can be calculated using (6.28) with $K = 3 \times 10^{10}$, $N = 10$, and $n = 32$ bit:

$$I_{cell} \sim 3 \cdot 10^{10} \cdot (\log_2 10 + 3 \cdot 32) \sim 3 \cdot 10^{12} bit \tag{6.30}$$

Expression (6.30) represents an upper bound on the information content that must be processed to assemble a new cell. This estimate can be refined in a number of ways [4]. For example, since the atomic composition, i.e. the frequency of occurrence of different elements, is known (Table 6.2), the Shannon Equation (6.23) can be used for a more accurate estimate of I_l (left to readers).

Earlier in this chapter, close relations between information and thermodynamic entropy of a system were briefly discussed. In fact, this relationship has allowed for experimental estimates of the information content of living cells based on microcalorimetric measurements. It has been concluded that the major consumption of energy during a cell's reproduction cycle arises from the correct placement of molecules within the cell [8]. The experimental information estimates for bacteria range from 10^{11}–10^{13} bits per cell [8]. Note that result (6.24) correlates well with experimental estimates. In the following, the conservative edge of the estimated range, i.e. I_{cell} ~ 10^{11} bits, is used.

6.3 ABSTRACT INFORMATION PROCESSORS

6.3.1 Turing machine

Alan Turing designed a paper model for universal computation which solves a broad spectrum of mathematical and logical problems in a finite number of steps [12]. This model, now called the *Turing machine,* is a hypothetical device that manipulates a *finite* set of symbols (in the simplest case '0's and '1's) according to a *finite* set of *rules*. The Turing machine consists of a line of symbols written on an infinite *tape* and a *monitor* or a read/write head with a finite number of internal states, i.e. finite automaton. The monitor reads a symbol from the input tape and consults its rule list. It then performs two actions: (i) it modifies the internal state of the monitor and (ii) it writes a symbol on the output tape (in the original design there was one input/output tape, later, machines with two and more tapes were proposed). A diagram of the Turing machine is shown in Figure 6.2a. Further developments of this concept led to the Universal Turing Machine that was a prototype of a general-purpose computer – a machine that could perform all possible numerical computations [13]. All practical computers have capabilities equivalent to the Universal Turing Machine.

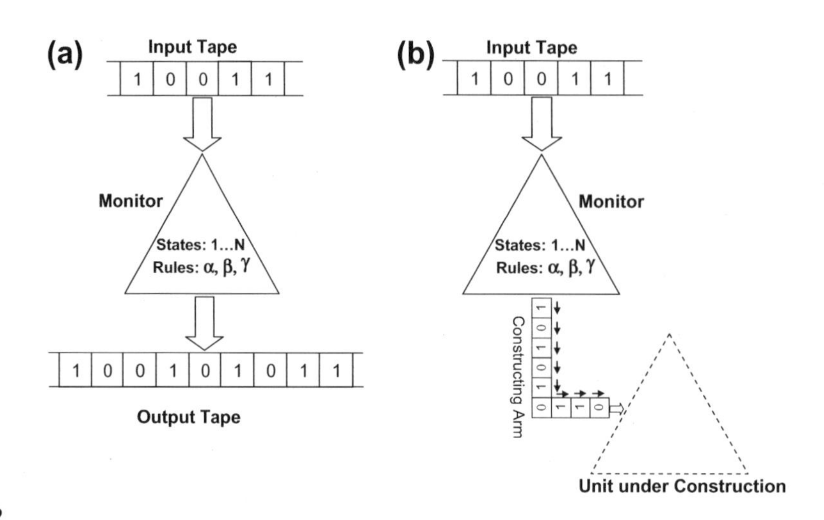

FIGURE 6.2

Abstract information processors: (a) Turing machine and (b) von Neumann universal constructor

6.3.2 Von Neumann universal automaton

John von Neumann further developed Turing's concept in application to practical computers. In particular, von Neumann was concerned with the question of complexity of automata that are capable of universal computation. Each automaton contains a certain number of discrete elements (e.g. transistors, resistors, diodes, etc.). The internal complexity of the system (i.e. the number of discrete elements K) defines the system capability. As von Neumann put it [14]: 'if one constructs the automaton (A) correctly, then any additional requirements about the automaton can be handled by sufficiently elaborated instructions. This is only true if A is sufficiently complicated, if it has reached a certain minimum of complexity'. In other words, a system cell must surpass a certain internal complexity threshold if it is to perform arbitrarily complex tasks by virtue of elaborate software instructions. This measure is sometimes called the *von Neumann threshold* [16], which is the smallest complexity of the system that could emulate general-purpose computing. Von Neumann hypothesized that the minimum circuit complexity required to implement general-purpose computing is on the order of a few hundred devices [15]. Recently the question of the minimal complexity of information processors was discussed in [16,17]. To estimate the von Neumann threshold, a 1-bit general-purpose computer was considered (referred to as the *Minimal Turing Machine*), which contained an arithmetic-logic unit and sufficient memory [9]. The minimum number of devices in the Minimum Turing Machine (MTM) was obtained in [9] to be $K_{MTM} = 314$, which is consistent with von Neumann's conjecture [15].

Von Neumann also worked on a general theory for complicated automata that would include both artificial machines, such as digital computers, and natural machines, such as living organisms [14]. Note that the Turing machine and related digital computers have inputs and outputs of completely different media than the automata themselves. Based on the generic nature of information, which is independent of the choice of its material carriers, von Neumann suggested the concept of a self-reproducing universal automaton, the *Universal Constructor*, which is the Turing machine expressing the output of computation in the same media as the automaton themselves: '…one imagines automata which can modify objects similar to themselves, or effect synthesis by picking up parts and putting them together, or take synthesized entities apart' [13,14]. A scheme of the von Neumann universal constructor is shown in Figure 6.2b. This revolutionary concept of computers making computers significantly expanded the horizons of the information universe. Among other things, it revealed the nature of manufacturing as an information-transfer process.

6.3.3 Living cell as a Turing machine

Living cells, such as bacteria, have the formal attributes of a Turing machine, i.e. a machine expressing a program [5,18]. In fact, the cell can be thought of as von Neumann's universal constructor, as the cell expresses the output of its information processing on the matter constituting the building blocks of the cell itself – thus *a computer making computers* [18]. In addition, single-cell organisms have been shown to exhibit the ability to learn, the ability to communicate with each other, various complex social behavior, etc.

The input information for the in carbo information processor comes from two sources: (i) from the cell's surroundings via a number of *sensors* called receptors, and (ii) by the cell's internal *memory* unit, the DNA molecule. The cell processes the input information via a series of biochemical processes, with two kinds of biomolecules playing a major role: RNA and proteins, which constitute the *logic* unit of the cell.

The output information of the in carbo processor has at least three components: (i) assembly of matter to make a new cell; (ii) actuation of motility organs, such as flagella, in response to external stimuli; and (iii) communication with other organisms.

The rest of this chapter deals with detailed characterization of the in carbo information processors in terms of the quantitative parameters of its units and with comparison between the in silico (nanomorphic cell) and in carbo systems.

6.4 IN SILICO AND IN CARBO SYSTEMS: A DESIGN PERSPECTIVE

6.4.1 In silico integration: Nanomorphic cell

As has already been discussed in Chapter 1, the 1000 μm^3 volume (e.g. 10 μm × 10 μm × 10 μm cube) of the nanomorphic cell must have the energy source, the control unit (includes logic and memory subunits), the communication system, and the sensor system. The goal of this section is to discuss the trade-offs that must be made in allocating volume resources for each of these subsystems. The work in the previous chapters will serve as a basis for volume allocations and the estimates of subsequent impacts on the performance of the nanomorphic cell.

Figure 6.3a shows a 'final' design of the nanomorphic cell as it has evolved throughout this book. It differs from the initial picture introduced in Chapter 1 (Fig. 1.3) in several ways:

1. The *control unit* in Figure 1.3 has been divided into *logic and memory units* in Figure 6.3.
2. The abstract multiparameter sensors in Figure 1.3 have been replaced by two specific sensors: a nanowire/nanotube FET (NWFET) bio(chemical) sensor and a thermocouple-based temperature

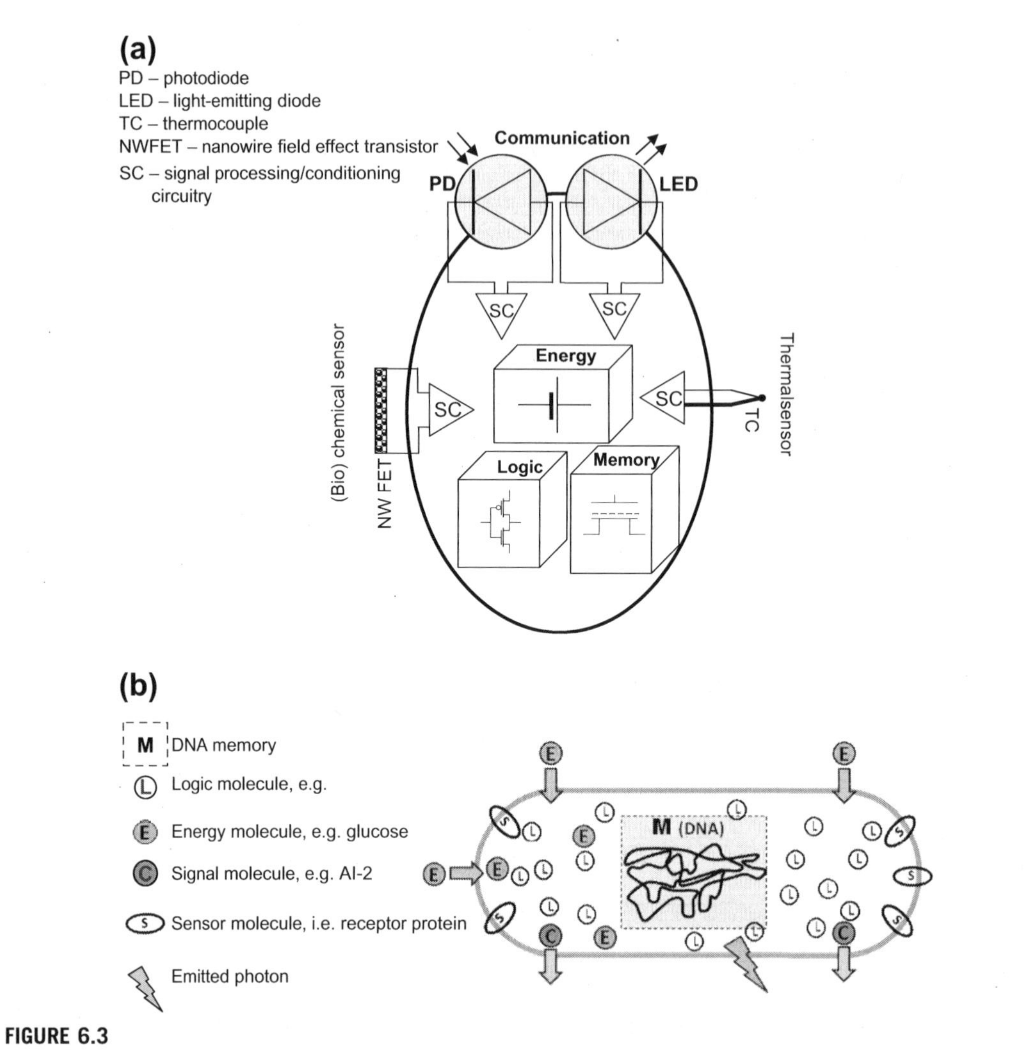

FIGURE 6.3

In silico (a) and in carbo (b) systems

sensor (TC). These choices for sensors for the nanomorphic cell are based on the discussion of several sensing options in Chapter 4.

3. The communication unit envisioned in Figure 1.3 as radio communication via an antenna to transduce electromagnetic radiation has been replaced by the infrared optical communication channel using a photodiode (PD) and a light-emitting diode (LED) to receive and transmit infrared signals respectively.
4. A galvanic cell has been selected for the energy source of the nanomorphic cell based on its superior volumetric energy and power characteristics, as discussed in Chapter 2.

Table 6.4 Volumetric parameters of the generic components of the nanomorphic cell

Component		Component volume	Volumetric density*
Energy		n/a	10^4 J/cm^3 = 10^{-8} J/μm^3
Control	Logic	$3a^3 = 3.8 \times 10^{-7}$ μm^{3**}	10^{17} bit/cm^3 = 10^5 bit/μm^3
	Memory	$45a^3 = 5.6 \times 10^{-6}$ μm^{3**}	10^{16} bit/cm^3 = 10^4 bit/μm^3
Sensing	Thermal	10 μm^3	n/a
	(Bio)chemical	2.5×10^{-4} μm^3	n/a
Communication	Transmitter	$(2\lambda)^3$ = 8 μm^{3***}	n/a
	Receiver	$(2\lambda)^3$ = 8 μm^{3***}	

*Gross value, taking into account device spacing and interconnects.
**a = 5 nm (projected minimal characteristic length in electronic devices).
***λ = 1 μm (infrared optical communication channel).

In the following, an allocation of volume for each of the units of the nanomorphic cell is offered based on the results of the limiting studies performed in Chapters 2–5 and tabulated in Table 6.4.

Note that some of the nanomorphic cell components are composed of relatively homogeneous collections of identical devices. For example, the memory and logic units are composed of a large number of devices, whose total number is proportional to the allocated volume. The capacity of the galvanic energy source is also proportional to its volume. Thus, logic, memory, and energy units have volumetric densities as performance parameters (Table 6.4).

On the other hand, the remaining primary components, e.g. sensors, the communication transmitter, and receiver, may be present as single components. They have fixed component volumes determined by the physics of their operation (Table 6.4).

Now, the available 1000 μm^3 is to be partitioned so as to provide adequate volume for all of the required components. Based on the data in Table 6.4, it appears that 100 μm^3 could provide sufficient volume to support all the 'singular' devices (leaving 900 μm^3 to be divided between energy, logic, and memory). The allocation strategy can be illustrated as follows. From Table 6.4, the total volume occupied by sensors, as well as by the communication transmitter and receiver, occupy a net volume slightly above 26 μm^3, thus leaving ~74 μm^3 for the corresponding signal processing/conditioning circuitry (SC), e.g. for signal amplification, filtering, analog-to-digital and digital-to-analog conversion, etc. Assuming equal volume for each of the four signal-processing units (i.e., the thermal and chemical sensors, transmitters and receiver) provides ~18 μm^3 per unit. This volume in principle allows for a reasonable complexity for the signal processing units, e.g. assuming a 3D circuit topology, the maximum number of transistors per signal processing unit is:

$$18\ \mu m^3 \times 10^5\ \text{bit}/\mu m^3 = 2 \times 10^6\ \text{(equivalent digital FET)}$$

(As was discussed in Chapter 4, the signal-conditioning circuits are constructed from devices similar to those in the logic unit, i.e. FET-type devices, which can operate both in analog and digital modes. In general, analog devices used in the signal-conditioning circuits are larger than the digital FET.)

With respect to the remaining 900 μm^3 of volume, suppose that one third of this volume or 300 μm^3 is allocated to each of these components. The overall count of the homogeneous components of a nanomorphic cell is:

Energy: $300\ \mu m^3 \times 10^{-8}\ J/\mu m^3 = 3 \times 10^{-6}\ J$
Logic: $300\ \mu m^3 \times 10^5\ bit/\mu m^3 = 3 \times 10^7\ bit$
Memory: $300\ \mu m^3 \times 10^4\ bit/\mu m^3 = 3 \times 10^6\ bit$

The computational capability of the nanomorphic cell that could be provided by the above levels of complexity is likely to be sufficient for the sense–analyze–announce function of the cell.

6.4.2 Introduction to in carbo systems design

Figure 6.3b presents an abstraction of a living bacterial cell (such as *E. coli*) as an in carbo functional microsystem. Similarly to the in silico microsystem of Figure 6.3a, it contains a DNA-based memory (M), cytoplasmic proteins implementing logic functions (L), membrane proteins acting as sensors (S), etc. The energy is supplied in chemical form via energy-storing molecules (E) such as glucose (glucose is sometimes referred to as *E. coli's* 'favorite dish'). Usually the internal storage of chemical energy is very limited in bacteria and therefore a constant supply of 'energetic molecules' from the outside is

Table 6.5 Essential parameters of *E. coli*

Macroparameters		Ref.
Cell length	2 μm	[9]
Cell diameter	0.8 μm (rod)	[9]
Cell total volume	$10^{-12}\ cm^3 = 1\ \mu m^3$	[9]
Cell surface area	$6 \times 10^{-8}\ cm^2 = 6\ \mu m^2$	[9]
Cell wet weight	10^{-12} g	[9]
Cell dry weight	3×10^{-13} g	[9]
Molecular processor		
Number of all RNA/cell	222 000	[9]
Number of cytoplasmic proteins	1 000 000	[9]
Number of ribosomal proteins	900 000	[9]
Number of all proteins	3 600 000	[9]
Time for cell replication	20–60 min*	[9,48,52,53]
Time for DNA replication	40 min	[48,53]
Energetics		
Number of glucose/cell	200 000–400 000	[9]
Number of ATP/cell	500 000– 3 000 000	[9]
Cell metabolic rate (power)	1.4×10^{-13} W	[11,52]

**A typical cell replication time of 40 min is used in all subsequent numerical analyses*

critical for survival of bacteria. To communicate with other cells the bacteria eject special signal molecules, which are detected by the recipient cells. Sometimes, cells can emit light, and there is some indication the process might also be used for interaction with the external world.

Data on the structure, composition, and operational parameters of the *E. coli* bacterium will be used to construct a benchmark in carbo system because the data for the *E. coli* are available with a great level of detail and accuracy (see Table 6.5). In the following sections, a snapshot discussion on each of the functional units of in carbo microsystem will be offered, i.e. *energy source*, *memory*, *logic*, *sensing*, and *communication*.

6.5 IN CARBO LONG-TERM MEMORY: STORING INFORMATION IN DNA

The in carbo memory unit (**M** in Fig. 6.3) contains information about the structure and operation of a living cell. The information is encoded digitally, in chemical form, by means of biomolecules, such as DNA and RNA. As indicated earlier, DNA coding uses the *base-4* (quaternary) system. Recall also that *base-4* has exactly the same information efficiency as *base-2* used in digital computing and is near the optimum base for computation. The four states are represented by four molecular fragments called *nucleobases* (or simply bases): *adenine* (**A**), *cytosine* (**C**), *guanine* (**G**), and *thymine* (**T**). The chemical structures of these molecular state-symbols are shown in Figure 6.3.

The four molecular state-symbols in DNA are attached in series to a flexible 'tape' built from sugar and phosphate groups, the whole structure and principle of information storage resembles magnetic tape storage [19]. The complete DNA molecule consists of two complementary 'tapes', forming the famous *double helix*. Each state-symbol on the first 'tape' forms a pair with a complementary state-symbol on the second 'tape': adenine (**A**) forms a pair with thymine (**T**), while cytosine (**C**) forms a pair with guanine (**G**), thus the complimentary base-pairs are **A–T** (**T–A**) and **G–C** (**C–G**). Information content in each 'tape' is identical but is written with different (complimentary) sequences of symbols.

The *base-pair* (bp) is a common unit of information stored in DNA (one bp corresponds to circa 3.4 Å of length along the 'tape'). Its relation to the binary information is set by (6.7) for $N = 4$ (the number of DNA state-symbols):

$$I(bp) = \log_2 4 = 2I(bit) \tag{6.31}$$

Thus, 1 bp = 2 bit.

Data on the DNA storage capacity (genome size) along with some macroparameters for several representative cellular organisms are given in Table 6.6. Note that the single-cell protozoan, *Amoeba dubia,* stores a huge amount of information (670 giga base-pairs or 1.34 Tbit) [20]) in its DNA. This is orders of magnitude larger than information stored in the human genome (~6 Gbit) and is the largest DNA storage capacity known for any organism.

6.6 IN CARBO LOGIC INFORMATION PROCESSION

When in silico electronic information processors are considered, such as a microprocessor or microcontroller, they are commonly referred to as logic units. They consist of a large number of

Table 6.6 DNA storage capacity (genome size) for several representative cellular organisms. Characteristic system macroparameters are provided for convenience

Cell	Volume, μm³	Size, μm	Power, W	Genome size
(Bacteria) – *E. coli*	1	0.8 × 2*	1.4×10^{-13}	4.6×10^6 bp = 9.6 Mbit
(Cyanobacteria) - *P. marinus*	0.1	0.5 × 0.7**	$\sim 10^{-15}$	1.75×10^6 bp = 3.5 Mbit
(Protozoa) - *Amoeba dubia*	$\sim 10^7$	~310***	$\sim 10^{-8}$	670×10^9 bp = 1.34 Tbit
Human cells****	~1000	~10	$\sim 4 \times 10^{-12}$	$\sim 3 \times 10^9$ bp = 6 Gbit

*Rod.
**Prolate spheroid.
***Length in locomotion.
****Average parameters.*

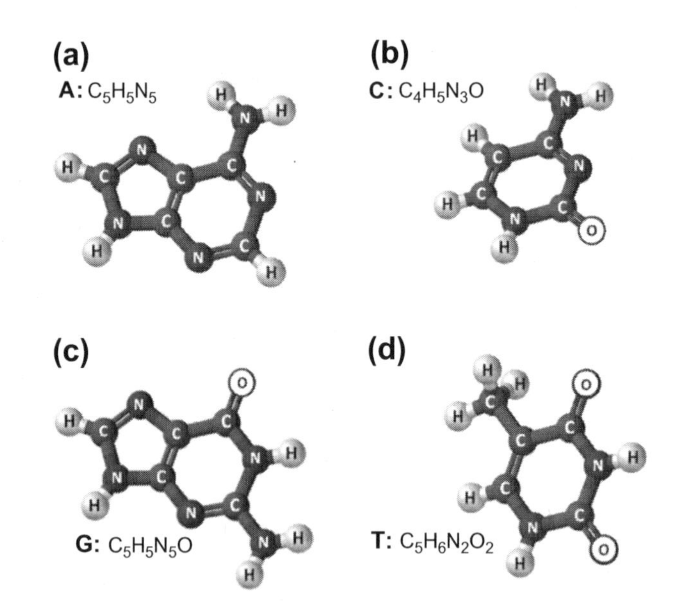

FIGURE 6.4

Chemical structure of four molecular state-symbols used for encoding information in DNA: (a) adenine; (b) cytosine; (c) guanine; and (d) thymine (the drawings are adapted from Wikipedia)

binary switches (transistors), which are generally viewed as logic devices (discussed in detail in Chapter 3). The total number of binary switches is a characteristic parameter of computational performance for a logic unit [16]. However, many binary switches inside of the logic unit are organized to perform short-term memory functions in the form of, e.g., registers and cache memories to support the logic operation. In fact these memory elements are necessary components of any general-purpose logic processor, including the minimal Turing machine [16]. If the

BOX 6.2 DNA MEMORY: READ AND WRITE OPERATIONS

READ: Different parts of the DNA memory of the cell are continuously accessed to support its operation. One example is *signal transduction*, which is the DNA-mediated/controlled process of cellular response to external stimuli. The outcome of signal transduction could be, e.g., change in metabolism, gene activation, alterations in cell's movements, etc. In fact, one initial stimulus can trigger a cascade of gene expressions, which can result in complex physiological events. The set of genes and the order in which they are activated in response to stimuli are often referred to as a *genetic program*. In fact, the DNA memory not only allows for continuous access to its different parts, it also acts as memory with multi-access capability by distinct computing units [47].

WRITE: The view that DNA is a read-only memory has undergone a dramatic change in recent years. Indeed (almost) exact copying of the parental DNA to the offspring, called *vertical gene transfer*, is the basis for inheritance and until recently was regarded as the only or at least vastly predominant mechanism for transferring the genetic information.

There is, however, an alternative mechanism for transfer of the genetic information, which is *lateral (horizontal) gene transfer* [48–51]. This can happen, for example, by a direct uptake ('swallowing') of naked DNA by a cell from its environment, by a virus, or by direct physical contact between donor and recipient cells. For example, fragments of the imported outside DNA can be integrated into the host DNA and thus new information is written in the memory unit. Until the advent of the genome-sequencing era, a prevailing opinion among the research community was that lateral gene transfer was a rare and insignificant event. Currently it is recognized that in prokaryotic, e.g. bacterial cells, lateral transfer is the predominant form of genetic variations and is one of the major driving forces for bacterial evolution. In fact, the scale of lateral gene transfer can be very large: for example two different strains of *E. coli*, which differ more radically in their genetic content than all mammals. The lateral gene transfer is also observed in eukaryotic cells, however its scale and role remains unclear.

memory elements built into a logic unit are organized in an array (e.g. cache memory in a microprocessor), a considerable portion of the logic circuitry is devoted to retrieving information from the memory array – thus serving as *memory interface* circuitry. All of the above memory-related elements are part of the logic unit in contrast to the more permanent and usually physically separate memory unit such as the magnetic hard disk drive or flash memory. If one applies a similar convention to the in carbo cellular information processor, the DNA component described in the previous section is described as the memory unit, while molecular elements used to retrieve the information from DNA and carry this information to certain locations in the cell can be regarded as part of the cell *logic*. Thus, in such a convention, the RNA molecules (see Box. 1.2) in Chapter 1 are counted as logic devices.

Many proteins in living cells have as their primary function the transfer and processing of information [21–24], and thus they also contribute to the logic devices count. Proteins can work as nano-devices by altering their 3D structural arrangement (conformation), and therefore they function according to the chemical–physical inputs from their microenvironments [23]. It should be noted that many proteins are used to build various macromolecular structures of the cell, while others participate in the cell's metabolism (the 'core function'). Such proteins should be excluded from the computational devices count [21]. However, as recent studies indicate, the proportion of components devoted to computational (signal) networks as opposed to those devoted to core functions increases with the complexity of the cell and are absolutely dominant in humans [23].

Some proteins and RNA in the cell are organized to form *ribosomes*. The function of ribosomes is to synthesize the protein molecules according to commands from DNA (provided by mRNA), and thus the ribosome can be regarded as a part of the cell's *logic unit* and, more exactly, its *output interface*.

An order-of-magnitude estimate for the number of logic devices in, e.g. *E. coli*, can be made based on the following considerations.

Logic device count includes *all* RNA and *some* protein molecules in the cell. According to Table 6.1, the total number of RNA molecules in the *E. coli* cell is

$$N_{RNA} = 2.2 \times 10^5 \tag{6.32a}$$

which can be regarded as a lower bound on the number of logic devices.

Also, the number of all proteins inside the cell is

$$N_{all\ protein} = 2.6 \times 10^6 \tag{6.32b}$$

Therefore, an estimated upper bound on the number of logic devices is the sum

$$N_{RNA} + N_{all\ protein} \sim 3 \times 10^6 \tag{6.32c}$$

On the other hand, according to the convention adapted in this section, all proteins in ribosomes are counted as part of the logic unit (output interface). The number of ribosomal proteins is:

$$N_{r\text{-}protein} = 9 \times 10^5 \tag{6.32d}$$

Therefore conservative estimate for the number of logic devices is

$$N_{logic} \sim N_{RNA} + N_{r\text{-}protein} \sim 10^6 \tag{6.32e}$$

A caveat needs to be mentioned when the number of 'logic' biomolecules in the in carbo systems is compared to that of in silico logic devices: While the in silico devices are typically binary switches (see Chapter 3), the in carbo devices can in principle represent multi-valued logic elements, as the number of different states in biomolecules (e.g. different 3D conformations of protein molecules) can be more than two.

6.7 IN CARBO SENSORS

Living cells need to get information about their environment in order to survive and reproduce. To do this they contain sophisticated sensing mechanisms to monitor their surroundings [24–26]. They can react to information about a range of external parameters, e.g. temperature, nutrients availability, dangerous chemicals, light, magnetic field, etc. The in carbo sensors (**S** in Fig. 6.3), called membrane/transmembrane receptors, usually are specialized proteins embedded in the cell's envelope. They react to different chemical and physical stimuli from the surroundings, e.g. by altering their 3D conformation. The receptor response in turn initiates a cascade of follow-up chemical events, known as *signal transduction*. Signal transduction can involve many agents and it eventually results in changes in the behavior of the cell.

The receptors usually form distinct clusters on the membrane surface that can include thousands of receptor proteins [27]. In *E. coli*, the receptor clusters are primarily localized at the cell poles [27].

BOX 6.3 DEATH RECEPTORS (DEATH SENSORS)

Perhaps the most intriguing types of cell sensors are *death receptors*, which participate in the cell's suicide mechanisms in multicellular organisms of animals, including humans. The cell suicide mechanism, called apoptosis, allows control of the cell number in tissues and elimination of those individual cells, which may threaten the survival of the living organism. Failure of a cell to commit suicide may result in cancer, while early suicide of the cells contributes to severe disorders such as Parkinson's and Alzheimer's diseases [28].

Certain cells have special sensors, which detect the presence of extracellular death signals and, in response, switch on the cell's internal suicide process [28].

6.8 IN CARBO COMMUNICATION

Single-cell organisms can send external signals to communicate with one another and even with higher organisms by several mechanisms, such as: (i) chemical communication via signal molecules; (ii) optical communication via bioluminescence; and (iii) 'tactile' communication via direct contact.

6.8.1 Chemical cell-to-cell communication

Single-cell organisms, such as bacteria, can communicate to each other by release and detection of special *signal molecules* [29–36]. The realization that bacteria communicate and alter their behavior accordingly, has led to a dramatic intensification of research in this area. In fact, in recent years there has been a paradigm shift in understanding of the unicellular world from the view of the non-cooperative bacterial cell to a new vision of complex social behavior of bacteria [33]. Cells use chemical signaling, e.g., to detect population density ('quorum sensing') and to exchange information about the local environment, the spatial distribution of the cells, etc. In general, chemical cell-to-cell communication coordinates the behavior of a cell population to increase access to nutrients, provide for collective defense, or enable the community to escape in case of threats to its survival [33]. Moreover, cross-interaction of microbes with higher organisms such as plants or mammals has also been reported [33].

The signaling molecules upon ejection from the sender cell move through the extracellular medium, e.g. by diffusion, and if met by another cell, attach to its membrane receptors, thus triggering changes in the function of the recipient cell. As a rule, the signal molecules are relatively small, which allows for sufficient mobility by diffusion in an aqueous environment. Two representative types of signaling molecules are *(oligo)peptides* (OP), which are short chains of 5–20 amino acids, and *N-acyl homoserine lactones* (AHL). While the oligopeptides and AHL signal molecules are specific to a given type of cell, a new type of molecule, termed *AI-2 (autoinducer 2)*, has recently been identified as a universal signaling molecule that is produced and sensed by a large number of different species [34]. The AI-2 molecules appear to be the universal 'language' for interspecies communication, the 'bacterial esperanto' [35]. Until recently, the chemical structure of AI-2 remained unknown, and only in 2002 [34], the structure of the AI-2 was determined to be *furanosyl borate diester* (molecular formula $C_5H_{10}O_7B$). The presence of boron in the signal molecule is very intriguing, as boron previously had no defined biological function [29]. Later, another derivative of AI-2, not containing boron, was also identified [36].

Table 6.7 Example signal molecules used in cell-to-cell communication: molecular weight and the metabolic energy cost of production

Signal molecules	Molecular formula	Molecular weight	Energy/molecule, E_{sig}
Oligopeptides (OP)	GluArgGlyMetThr ($C_{22}H_{48}N_8O_{13}S$)*	664	~10^{-17} J (184 ATP)
N-acyl homoserine lactones (AHL)	$C_8H_{12}O_4N$**	188	~4×10^{-18} J (8 ATP)
Autoinducer-2 (AI-2)	$C_5H_{10}O_7B$***	193	~5×10^{-20} J (1 ATP)

*CSF (Bacillus subtilis).
**LuxLM (Vibrio Harvey).
***AI-2 (Vibrio Harvey).

Energy costs of communication

The metabolic energy cost of production of different signal molecules was estimated in [31] and is shown in Table 6.7. The total energy of communication depends on the number of signal molecules involved, which in turn depends on the communication distance (also called cell-to-cell 'calling distance'), as is discussed below.

Communication distance

Chemical signaling is used for short-distance communication, typically <10 μm, although longer communication distances up to 78 μm have been observed [30]. A simple model for estimates of the limits of chemical communication is presented below.

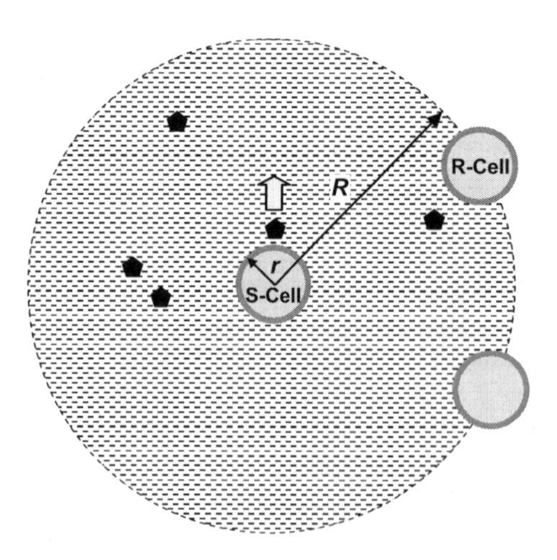

FIGURE 6.5

Cell-to-cell (bio)chemical communication

Consider a nearly spherical sender cell with a 'radius' $r \sim 5$ µm. Suppose the sender cell expels N_{sig} signal molecules to its exterior and a neighbor recipient cell is located at distance R from the sender (Fig. 6.5). As was discussed in Chapter 4 (Box 4.4), the resulting molar concentration within a spherical volume of radius R is

$$\mu_{sig} = \frac{N_{sig}}{N_A} \bullet \frac{3}{4\pi\left(R^3 - r^3\right)} \tag{6.33a}$$

Alternatively, the number of molecules for a given concentration is:

$$N_{sig} = \frac{4}{3}\pi\left(R^3 - r^3\right) \bullet N_A \bullet \mu_{sig} \tag{6.33b}$$

The total energy of communication is the product of the number of ejected signal molecules and the energy cost E_{sig} of production of each signal molecule (see Table 6.6).

$$E_{com} = N_{sig} \bullet E_{sig} \tag{6.33c}$$

The minimal concentration of signal molecules detectable by a recipient cell was found to be in the range $\mu_{sig} = 1$–10 nM [32]. Assuming $\mu_{sig} = 5$ nM, i.e in the middle of the range, the required N_{sig} (6.33b) and thus the energy of communication E_{com} (6.33c) can be calculated as a function of distance, R, for the three categories of typical signal molecules listed in Table 6.6. The plot of required energy for communication as a function of transmission distance is shown in Figure 6.6.

As can be seen in Figure 6.6, in carbo communication is an energy-costly process as it also is for in silico systems (discussed in Chapter 5). In fact, for signal molecules OP or AHL, if the communication distance reaches ~ 200 µm, then the required communication energy approaches the total energy that the cell (e.g. *E. coli*) uses during the entire reproduction cycle, $E_{com} \approx E_{cell} \approx 3 \times 10^{-10}$ J (see Section 6.9 below for a discussion of energy properties of the cell). If a communication time $t_{com} \sim 1$s is assumed, then the communication power $P_{com} = E_{com}/t_{com}$ can exceed the total operational power of the cell (e.g. $\sim 10^{-13}$ W for *E. coli*) at distances ~ 100 µm, even if low-energy signal molecules AI-2 are used.

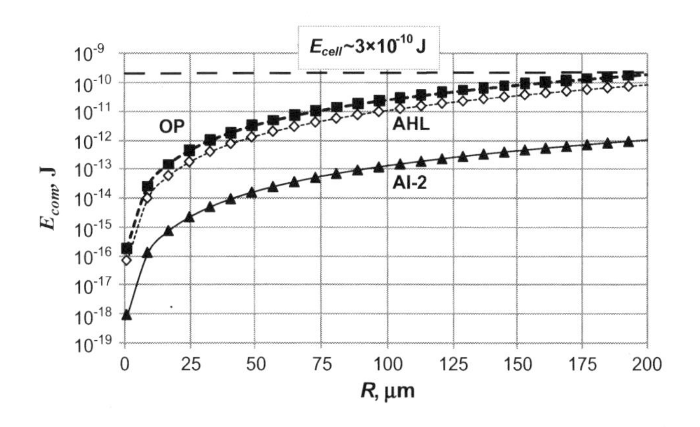

FIGURE 6.6

Communication energy costs vs. distance for three types of signal molecules.

6.8.2 Optical signaling/communication

Many unicellular organisms are able to emit light [37–39]. The luminescence typically occurs as a result of a biochemical reaction catalyzed by special proteins, generally referred to as *luciferase*. A typical wavelength of emitted light is $\lambda \sim 500$ nm and thus the energy of emitted photons (see Chapter 5) is $E_{ph} \sim 4 \times 10^{-19}$ J or ~2.5 eV. As has already been discussed in Chapter 5, the photon emission process is very energy costly. Indeed, by emitting light a bacterial cell can consume up to 20% of its energy, which is a very energy costly expenditure for bacteria [38].

The biological role of the bioluminescence remains unclear. If the light bacteria form a symbiosis with a higher organism (e.g. bacterial light organ in many fish), this host organism can use the light emitted by bacteria to attract prey, escape from predators, or for communication [38]. In return for light production, the bacteria are provided with nutrients, protection, etc.

Luminescence of free-living bacteria poses even more questions, as the energy cost of the process is very high and is difficult to justify. One view is that there is a positive selection and that the selection mechanism may relate to the dispersion and propagation of the bacteria [37]. For example, it can be speculated that for some marine bacteria, the attraction of organisms that will ingest them is advantageous [37].

The cell-to-cell communication function of light emitted by cells is not proven. It should be noted, however, that luminescence in most bacteria is under cell-density-dependent regulation and is triggered by the 'quorum-sensing' process discussed above. In fact, the discovery of chemical communication via signal molecules stemmed from the studies of bacterial luminescence, which led to the notion of quorum sensing [39].

There is a school of thought that some bacteria and other living cells may have a direct mechanism of communication via the emission and detection of light, the phenomenon referred to as *biophoton emission* [40, 41]. Biophoton emission differs from classical bioluminescence both in light intensity, i.e. ~10^4 photons/s per cell in *bioluminescence* [37] vs. ~10 photons/s per cell in *biophoton emission* [40], and wavelength, i.e. *bioluminescence* typically refers to visible light (e.g. $\lambda \sim 500$ nm [39]) while *biophoton emission* often refers to ultraviolet range (e.g. $\lambda \sim 300$ nm [40]). There are studies suggesting that biophotons act as transmitters of information [40, 41]. However, the research community has yet to reach a consensus on the nature and biological role of radiation by living cells. One of the difficulties associated with quantitative research in this field is the low level of optical signals generated by bacteria relative to the environmental photon background.

6.8.3 Direct contact communication

Some single-cell organisms, such as *myxobacteria*, communicate with each other by direct cell contact (so-called *C-signaling* system) [42]. Such tactile communication probably utilizes the least energy of all cellular communication mechanisms.

6.9 IN CARBO ENERGY SOURCE

Bacteria such as *E. coli* depend on a constant energy supply from their environment. The favorite food of *E. coli* is glucose. However, the reserves of glucose that are typically stored inside the cell are very limited (see Table 6.5), and therefore harvesting of chemical energy from outside of the cell is mandatory. For example, the *E. coli* cell contains up to 400 000 glucose molecules (Table 6.5) and each glucose molecule can produce ~30 eV ~ 5×10^{-18} J of energy (see Table 2.2.2 in Chapter 2). Thus, based on Table 6.5, the total energy stored in cellular glucose in an *E. coli* cell is ~2×10^{-12} J. Note that in addition to glucose, energy is also stored in a number of derivatives of the glucose catabolism process; therefore the total stored energy is somewhat larger, though it remains in the range estimated above for glucose.

On the other hand, the power consumption of *E. coli* is about 1.4×10^{-13} W (Table 6.5) and the cell divides approximately every 2400 seconds (40 minutes – see footnote for Table 6.5). Thus the total energy the cell uses during one division cycle is

$$E_{cell} = 1.4 \times 10^{-13}\ \text{W} \times 2400\ \text{s} \approx 3 \times 10^{-10}\ \text{J} \tag{6.34}$$

When starving, *E. coli* consumes some of its ribosomes as an internal source of both energy and nutrients (e.g. carbon and nitrogen) [43]. In fact, starvation of a population of *E. coli* cells results in disintegration of some of the cells (*lysis*) within a few hours. Those cells that remain alive are then able to utilize the remnant of dead bacteria to support their existence [44].

While at typical conditions occurring in nature *E. coli* does not have significant internal energy storage, there are also exceptions. At some abnormal conditions *E. coli* can synthesize as much as 20% of its dry weight as glycogen (glucose polymer) [54], which corresponds to ~10^{-9} J of stored energy. This amount of energy is equivalent to ~3x the cell cycle energy budget (6.34).

Many bacteria do not survive under long deprivation of nutrients. However, there can be remarkable exceptions in which some bacteria have developed amazing mechanisms and optimization strategies to live under nutrient limitations [44–46]. Also cells can utilize other sources of energy such as light as was briefly discussed in Chapter 1. Eukaryotic cells, on the other hand, which are generally larger and more complex than bacteria, can store a considerable amount of energy internally.

6.10 BENCHMARK IN CARBO INFORMATION PROCESSOR

6.10.1 Top-down estimate of overall computational performance

As was discussed above in Section 6.2.6, a conservative edge for estimated information content of a bacterial cell such as *E. coli* is I_{cell} ~ 10^{11} bits. In other words, for the correct placement of all molecules within the cells, e.g. by the von Neumann's universal constructor, 10^{11} bits needs to be generated by the cell processor. This quantity will be referred below as the binary information content per task, $I_{task} = I_{cell}$, where 'task' refers to a completed assembly of a cell.

A measure of the information-processing rate can be made for the time required for a cell to reproduce itself. A typical reproduction time of an *E. coli* cell is 2400 seconds (Table 6.5), and this is taken to be time required for one computational task/cycle, t_{task}. Thus the number of bits that must be processed per second (bit rate), $F_{in\ carbo}$, by the *E. coli* cell is:

$$F_{in-carbo} = \frac{I_{task}}{t_{task}} \sim \frac{10^{11}}{2400} \approx 10^7\ \text{bit/s} \tag{6.35}$$

It is assumed that each operation of the in carbo logic 'device' processes one bit, then $F_{in\ carbo}$ can be interpreted as having units of switching events per second. The power consumption of *E. coli* is about 1.4×10^{-13} W (Table 6.6) so that

$$P = F_{in-carbo} \cdot E_{bit} \tag{6.36}$$

where E_{bit} is the energy per equivalent binary operation, and from (6.35) and (6.36), the energy per bit operation in the in carbo information processor can be estimated as:

$$E_{bit} = \frac{P_{cell}}{F_{in-carbo}} = \frac{1.4 \cdot 10^{-13}}{10^7} \approx 10^{-20}\ \text{J/bit} \approx 10\ k_B T \tag{6.37}$$

Note that since there are approximately $N \sim 10^6$ logic devices in *E. coli* (see Section 6.6), the average frequency at which each device operates can be estimated as:

$$f \sim \frac{F_{in-carbo}}{N} \sim \frac{10^7}{10^6} \sim 10Hz \tag{6.38}$$

Finally, it should be noted that the energy utilization per switching event given by (6.37) may be quite conservative. The total number of equivalent binary operations needed to generate output information of 10^{11} bits may be greater than 10^{11}. Nevertheless even at the conservative edge, the estimated energy utilization per switching event is quite impressive. This number can be compared to the best-case scenario for in silico processors (Table 3.7 in Chapter 3):

$$\text{In silico: } 10^{-19} - 10^{-18}\ \text{J}$$

$$\text{In carbo: } 10^{-20}\ \text{J}$$

As can be seen the energy per bit operation in the in carbo systems is at least one or two orders of magnitude lower than the most optimistic projections for in silico systems.

6.10.2 Bottom-up look: Memory and logic of in carbo vs. in silico

Memory

As was discussed in preceding Section 6.5, the in carbo processor of *E. coli* contains ~10 Mbit of DNA memory (Table 6.6), all confined within a 1 μm^3 volume, and this results in a 3-dimensional memory density of ~10^{19} bit/cm^3 (assuming the memory is uniformly distributed throughout the volume). In contrast, the 3D density of the 'minimal electronic memory' is ~10^{16} cm^{-3}, as was derived in Chapter 3.

Logic

In Section 6.6, the number of logic elements in *E. coli* was estimated to be ~10^6, thus implying 10^{18} cm^{-3} device density. In contrast minimum size electronic logic elements can be packed in a 3D volume with a maximal density of ~10^{17} cm^{-3} as discussed in Chapter 3. A summary for the in silico/in carbo comparison is given in Table 6.8 for 1 μm and 10 μm systems respectively (the data for 10 μm *in carbo* system is obtained by scaling of the 1 μm *E. coli.* cell data).

It is clear from Table 6.8 that the in silico system is less dense in the limits of scaling and that it is less energy efficient at the device level than a corresponding in carbo system. In the following, a few thoughts are offered as a possible explanation for these significant differences in performance.

Table 6.8 Comparison of element properties of the in silico and in carbo systems for 1 μm and 10 μm overall dimensions

1-μm system	In silico*	In carbo
Memory	~10^3 bit	~10^7 bit
Logic	~10^4 bit	~10^6 bit
Energy per bit	~10^{-18} J	~10^{-20}J
Power	10^{-8} W (max available)	~10^{-13} W (average used)
Heat flux	0.17 W/cm^2	~10^{-6} W/cm^2
10-μm system	**In silico***	**In carbo**
Memory	3 × 10^6 bit	~10^{10} bit
Logic	3 × 10^7 bit	~10^9 bit
Energy per bit	~10^{-18} J	~10^{-20}J
Power	10^{-6} W (max available)	~10^{-10} W (average used)
Heat flux	0.17 W/cm^2	~10^{-5} W/cm^2

See Section 6.4.1 for a detailed discussion.

6.10.3 Power and heat dissipation

The maximum power available for the nanomorphic cell can be estimated using (2.2.5b) from Chapter 2 and the energy source space allocation discussed in Section 6.4.1. The result is that the maximum available power $P \sim 10^{-8}$ W and $P \sim 10^{-6}$ W respectively for the overall dimensions $l = 1$ μm and $l = 10$ μm. The surface area of the nanomorphic cell is $A = 6l^2$ (cubic shape assumed) and therefore the maximum heat flux, Q, generated by the nanomorphic cell is

$$Q = \frac{P}{6l^2} \tag{6.39}$$

The values for the nanomorphic cell power and dissipated heat are given in Table 6.8. Note that the heat density generated by the nanomorphic cell is always below 1 W/cm^2, and this amount of heat is manageable by conventional techniques (see Chapter 2 for an in-depth discussion of heat removal). Also, from Table 6.8 the heat flux for the in carbo cell is at least four orders of magnitude lower than that for the in silico cell.

6. 10.4 Design secrets of an in carbo system

Heavier mass of information carrier

As it was argued in Chapter 3, a heavier mass of the information carrier allows for smaller separation between distinguishable states and therefore more states per unit volume or area. The characteristic dimension, a, as given by the Heisenberg relation (see Chapter 3 for details) decreases as the mass of the information carrier increases:

$$a \sim \frac{\hbar}{2\sqrt{2mE_b}} \tag{6.40}$$

For example, DNA memory uses molecular fragments (nucleotides) as information carriers, each consisting of more than 10 atoms. The molecular information carriers are densely packed in a linear array with distance between nucleotides of only 0.34 nm [19]. By comparison, the critical dimension of electron memory is ~5 nm, i.e. more than ten times more that of DNA. This explains the 1000× difference in volumetric memory density between electronic and DNA memory, i.e. 10^{16} bit/cm^3 of electronic memory vs. 10^{19} bit/cm^3 for DNA memory.

Utilization of ambient thermal energy

For in silico systems thermal energy (~ k_BT) must be managed as it may destroy the state or divert the information carrier from its intended trajectory; for example in communication between several logic elements. In order to overcome the deleterious effects of thermal energy each logic or memory element must contain a barrier $E_b > k_BT$. Moreover, in the communication with other elements, N carriers must be sent to the recipient elements, each of which must have kinetic energy $E_k > k_BT$. As a result the total energy of device operation, as it was derived in Chapter 3, becomes

$$E_{SW} = 2E_b + N \cdot E_k \tag{6.41}$$

and it can be significantly large, usually $>100\ k_BT$.

In contrast, the in carbo systems utilize thermal energy to effect data exchange/transmission between, e.g., logic-to-logic or memory-to-logic elements. All computational molecules move within the cell's space by thermally excited random walk with no extra energy required, and thus the second term in (6.41) is eliminated. In carbo systems actually uses thermal energy in the transmission of information!

Flexible/on-demand 3D connections/routing

Once more, referring to Eq. (6.41), in in silico systems most energy is consumed by interconnect. This is due to the need to pump a large number, N, of carriers (electrons) into the interconnecting wire for reliable communications. As was shown in Chapter 3, for reliable communication, N must dramatically increase for longer path lengths and/or more receiving devices (fan out). In electrical circuits the connection paths are pre-determined and in many instances, the electron travels a long distance from, e.g., point **A** to point **B**. Devices of in carbo systems are usually free to travel in all three dimensions within the cell and they don't follow a fixed path.

6.11 SUMMARY

In this chapter, the essential units of the nanomorphic cell (energy, control, communication and sensing) were combined within the 1000 μm^3 volume (e.g.10μm×10μm×10μm cube). The corresponding trade-offs that must be made in allocating volume resources for each of these units were discussed. It was concluded that the computational capability of the nanomorphic cell that could be sufficient to enable the sense-analyze-announce function of the cell.

Also, in this chapter an effort has been made to compare the projected performance of the nanomorphic cell (in silico system) with that of the living cell (in carbo system). The approach was to adopt the view of the living cell as a 'universal constructor', a type of computer that makes copies of itself, and that was first suggested by von Neumann. In order to provide a common framework for comparisons, it was necessary to introduce a few concepts from mathematical information theory. One interesting observation from this study was that the base-2 information system, prevalent in digital computation and base-4 system used by biological system are equally efficient numerical/symbolic representations.

To develop quantitative comparisons between the in silico and in carbo systems, it was necessary to estimate the size of permanent memory and the number of logic-processing units in the living cell. The *E. coli* cell, which has been widely studied, was used to develop estimates for the number of each of these elements and to bound the amount of information that must be generated by the cell to effect reproduction. There exist estimates of the power consumption of various living cells and these data were used to develop estimates of bounds for the energy consumption of the 'computational components' of the living cell, and for their effective rates of information processing.

Living cells can communicate/socialize with neighboring cells by various means including biochemical molecule emission, direct contact, and possibly optical. In relation to the communication schemes of nanomorphic cell, it occurs that cell-to-cell communication in both in silico and in carbo systems requires a significant and similar expenditure of energy.

The comparisons offered in this chapter suggest that the amazingly complex living cell is unexcelled in the efficient use of limited energy sources to perform the computation needed to support its existence. It has also been argued that the living cell is demonstrably superior in device density and energy consumption to the most optimistically scaled electronic cell of comparable dimensions. Further careful analyses are needed to refine the estimates offered herein but the underlying message is that practitioners of energy-efficient inorganic computation may find the computational strategies of the living cell to be a source of inspiration for more efficient computation.

APPENDIX: CHOICE OF PROBABILITY VALUES TO MAXIMIZE THE ENTROPY FUNCTION

Let

$$J = -C\sum_{i=1}^{K} p_i \ln(p_i) + \lambda\left(\sum_{j=1}^{K} p_i - 1\right) \tag{A1}$$

In (A1), C is an arbitrary constant and λ is a Lagrange multiplier enforcing the constraint that the sum of probabilities must be one. It is necessary that the derivative of *J* with respect to each probability be equal to zero. Therefore,

$$\frac{\partial J}{\partial p_i} = -C[\ln(p_i) + 1] + \lambda = 0; \; i = 1, 2, \ldots, K \tag{A2}$$

It follows that

$$p_i = e^{\frac{\lambda}{C} - 1}; \; i = 1, 2, \ldots, K \tag{A3}$$

Clearly from (A3), each of the probabilities is equal to the same constant and we must have that the sum of the probabilities is one, i.e.,

$$\sum_{i=1}^{K} p_i = 1 = Ke^{\frac{\lambda}{C}-1} \tag{A4}$$

(A4) can be solved for λ to yield

$$\lambda = C\left[1 + \ln\left(\frac{1}{K}\right)\right] \tag{A5}$$

The substitution of (A5) into (A3) yields the choice for probabilities that maximize (A1):

$$p_i^* = e^{\left(\frac{C[1+\ln(1/K)]}{C} - 1\right)} = e^{\ln\left(\frac{1}{K}\right)} = \frac{1}{K}; \quad i = 1, 2, \ldots, K \tag{A6}$$

The maximum value of the information entropy that results is

$$-C\sum_{i=1}^{K} p_i^* \ln(p_i^*) = -C\sum_{i=1}^{K} \frac{1}{K}\ln\left(\frac{1}{K}\right) = C\ln(K) \tag{A7}$$

LIST OF SYMBOLS

Symbol	Meaning
A	Area
b	Computational base, logarithm base
C	Constant
E	Energy
f	Frequency
F	Bit rate
I	Information
k_B	Boltzmann constant, $k_B = 1.38 \times 10^{-23}$ J/K
K, L, N	Integer numbers
l	Length
m	Mass
M	Molar mass
N_A	Avogadro's Number, $N_A = 6.022 \times 10^{23}$ mol^{-1}
p	Probability
P	Power
Q	Heat flux
r, R	Radius
S	Entropy

Symbol	Meaning
t	Time
λ	Wavelength
μ	Molar concentration
ψ	Information efficiency function
$\propto$	Indicates proportionality
$\sim$	Indicates order of magnitude

References

[1] T.P. de Souza, P. Stano, P.L. Luisi, The minimal size of liposome-based model cells brings about a remarkably enhanced entrapment and protein synthesis, ChemBioChem 10 (2009) 1056–1063.

[2] L. Brillouin, Science and Information Theory, Academic Press, New York, 1962.

[3] A.M. Yaglom, I.M. Yaglom, Probability and Information, D. Reidel, Boston, 1983.

[4] R.U. Ayres, Information, Entropy and Progress, AIP Press, New York, 1994.

[5] O. B-Küppers, Information and the Origin of Life, The MIT Press, 1990.

[6] H.R. Horton, L.A. Moran, R.S. Ochs, J.D. Rawn, K.G. Scrimgeour, Principles of Biochemistry, Prentice-Hall, Inc, 1996.

[7] E. Smith, Thermodynamics of natural selection I: Energy flow and the limits on organization, J. Theoret. Biol. 252 (2008) 185–187.

[8] W.W. Forrest, Entropy of microbial growth, Nature 225 (1970) 1165–1166.

[9] *E. coli* Statistics, University of Alberta, http://gchelpdesk.ualberta.ca/CCDB/cgi-bin/STAT_NEW.cgi

[10] The Microbial World, University of Wisconsin – Madison, http://textbookofbacteriology.net/themicrobialworld/nutgro.html

[11] A.M. Makarieva, V.G. Gorshkov, B.-L. Li, Energetics of the smallest: do bacteria breathe at the same rate as whales? Proc. R. Soc B 272 (2005) 2219–2224 (ELECTRONIC APPENDIX).

[12] A.M. Turing, On computable numbers, with an application to the Entscheidungsproblem, Proc. Lond. Math. Soc. 42 (1936) 230–265; A.M. Turing, On computable numbers, with an application to the Entscheidungsproblem - A correction, Proc. Lond. Math. Soc. 43 (1937) 544–546.

[13] N.G. Cooper, From Turing and von Neumann to the present, Los Alamos Science (Fall 1983) 22–27.

[14] J. von Neumann, Theory of Self-Reproducing Automata, Univ. of Illinois Press, 1966.

[15] J. von Neumann, The Computer and the Brain, Yale Univ. Press, 1959.

[16] V.V. Zhirnov, R.K. Cavin, Scaling beyond CMOS: Turing-Heisenberg rapprochement, Solid-State Electron 54 (2010) 810–817.

[17] V. Zhirnov, R. Cavin, G. Leeming, C. Galatsis, Assessment of integrated digital cellular automata architectures, COMPUTER 41 (2008) 38–44.

[18] A. Danchin, Bacteria as computer making computers, FEMS Microbiol. Rev. 33 (2009) 3–26.

[19] G. Bate, Bits and Genes: A comparison of the natural storage of information in DNA and digital magnetic recording, IEEE Trans. Magn. 14 (1978) 964–965.

[20] L.W. Parfrey, D.J.G. Lahr, L.A. Katz, The Dynamic Nature of Eukaryotic Genomes, Molecular Biol. Evolut. 25 (2008) 787–794.

[21] D. Bray, Protein molecules as computational elements in living cells, Nature 376 (1995) 307–312.

[22] N. Ramakrishnan, U.S. Bhalla, J.J. Tyson, Computing with proteins, Computer 42 (2009) 47–56.

[23] L.F. Agnati, D. Guidolin, C. Carone, M. Dam, S. Genedani, K. Fuxe, Understanding neuronal cellular network architecture, Brain Res. Rev. 58 (2008) 379–399.
[24] A. Wagner, From bit to it: How a complex metabolic network transforms information into living matter, BMC Systems Biology 1 (2007) 33.
[25] I.D. Campbell, The Croonian lecture 2006: Structure of the living cell, *Phil. Trans. R. Soc.* B (2008).
[26] J.G. Mitchell, The energetics and scaling of search strategies in bacteria, Amer. Naturalist 160 (2002) 727–740.
[27] D.M. Morris, G.J. Jensen, Toward a biomechanical understanding of whole bacterial cells, Annu. Rev. Biochem. B 363 (2008) 2379–2391.
[28] A. Ashkenazi, V.M. Dixit, Death receptors: signaling and modulation, Science 281 (1998) 1305.
[29] B.L. Bassler, Small talk: Cell-to-cell communication in bacteria, Cell 109 (2002) 421–424.
[30] A.W. Decho, R.S. Norman, P.T. Visscher, Quorum sensing in natural environments: emerging views from microbial mats, Trends in Microbiol. 18 (2010) 73–80.
[31] L. Keller, M.G. Surette, Communication in bacteria: an ecological and evolutionary perspective, Nature Reviews 4 (2006) 249–258.
[32] J. Müller, C. Kuttler, B.A. Hense, Sensitivity of the quorum sensing system is achieved by low pass filtering, BioSystems 92 (2008) 76–81.
[33] P. Williams, Quorum sensing, communication and cross-kingdom signaling in the bacterial world, Microbiology 153 (2007) 3923–3938.
[34] X. Chen, S. Schauder, N. Potter, A. Van Dorsselaer, I. Pelczer, B.L. Bassler, et al., Structural identification of a bacterial quorum sensing signal containing boron, Nature 415 (2002) 545–549.
[35] C. Fuqua, E.P. Greenberg, Listening in on bacteria: acyl-homoserine lactone signaling, Nature Rev. 3 (2002) 685–695.
[36] S.T. Miller, K.B. Xavier, S.R. Campagna, M.E. Taga, M.F. Semmelhack, B.L. Bassler, et al., *Salmonella typhimurium* recognizes a chemically distinct form of the bacterial quorum-sensing signal Al-2, Molecular Cell 15 (2004) 677–687.
[37] K.H. Nealson, J.W. Hastings, Bacterial bioluminescence: Its control and ecological significance, Microbiol. Rev. 43 (1979) 496–518.
[38 G. Wegrzyn, A. Czyż, How do marine bacteria produce light, why are they luminescent, and can we employ bacterial luminescence in aquatic biotechnology? Oceanologia 44 (2002) 291–305.
[39] T. Wilson, J.W. Hastings, Bioluminescence, Annu. Rev. Cell Dev. Biol. 14 (1998) 197–230.
[40] Y.A. Nikolaev, Distant interactions in bacteria, Microbiology 69 (2000) 597–605.
[41] D. Fels, Cellular communication through light, PLOS ONE 4 (2009) 1–8.
[42] C.J. Tomlin, J.D. Axelrod, Biology by numbers: mathematical modeling in developmental biology, Nature Reviews 8 (2007) 331–340.
[43] A.L. Koch, What size should a bacterium be? A Question of scale, Annu. Rev. Microbiol. 50 (1996) 317–348.
[44] J.C. Ensign, Long-term starvation survival of rod and spherical cells of *Arthrobacter crystallopoietes*, J. Bacteriology 103 (1970) 569–577.
[45] B. Velimirov, Nanobacteria, ultramicrobacteria and starvation forms: A search for the smallest metabolizing bacterium, Microbes and Environments 16 (2001) 67–77.
[46] R. Cavichioli, M. Ostrovski, F. Fegatella, A. Goodchild, N. Guixa-Boixereu, Life under nutrient limitation in oligotrophic marine environments: An eco/physiological perspective of Sphingopyxis alaskensis, Microbial Ecology 45 (2003) 203–217.
[47] D.J. D'Onofrio, G. An, A comparative approach for the investigation of biological information processing: An examination of the structure and function of computer hard drives and DNA, Theor. Biol. And Med. Model. 7 (2010) 3–20.

[48] N. Lane, Mitochondria: Key to complexity, in: W. Martin, M. Müller (Eds.), Origin of Mitochondria and Hydrogenosomes, Springer-Verlag Berlin, Heidelberg, 2007.

[49] H. Ochman, J.G. Lawrence, E.A. Groisman, Lateral gene transfer and the nature of bacterial innovation, Nature 405 (2000) 299–304.

[50] E.V. Koonin, K.S. Makarova, L. Aravind, Horizontal gene transfer in prokaryotes: quantification and classification, Annu. Rev. Micribiol. 55 (2001) 709–742.

[51] C.M. Thomas, K.M. Nielsen, Mechanisms of, and barriers to, horizontal gene transfer between bacteria, Nature Reviews 3 (2005) 711–721.

[52] J.P. DeLong, J.G. Okie, M.E. Moses, R.M. Sibly, J.H. Brown, Shifts in metabolic scaling, production, and efficiency across major evolutionary transitions of life, PNAS (2010), in press (published online before print June 29, 2010): http://www.pnas.org/content/early/2010/06/25/1007783107.abstract

[53] S. Cooper, C.E. Helmstetter, Chromosome replication and the division cycle of *Escherichia coli* B/r, J. Mol. Biol. 31 (1968) 519–540.

[54] J.F. Wilkinson, Carbon and energy storage in bacteria, J. Gen. Microbiol. 32 (1963) 171–176.

Concluding remarks

In this book, the thought-problem of designing a highly functional, micron-scale electronic system has been explored in the context of possible in vivo applications in the human body. The idea of designing this nanomorphic system, although inspired by the continued scaling of feature sizes of transistors and memory elements in integrated circuit technology, extends the scaling concept in at least two ways. First, additional components of the nanomorphic cell such as energy sources, sensors and communication systems must also be scaled into micron or submicron dimensions. This property of the nanomorphic cell has a relation to system-on-a-chip technology but with an increased emphasis on the variety of components that must be scaled and assembled into an integrated system. A second differentiation from classical integrated circuit technology is the physical size of the nanomorphic system and the implied fabrication technology. In the integrated circuit case, the systems have dimensions on the order of centimeters; much larger than the micron-scale dimensions of the nanomorphic cell.

One of the encouraging findings in the book is the degree of support for realization of the nanomorphic cell provided by the continued scaling of logic and memory elements. It appears that thousands of logic and memory elements could be integrated into the 10 micron nanomorphic cell if scaled device feature sizes in the far submicron regime are achieved by the industry as planned. Such devices are projected to be available in the 2020 timeframe by the International Technology Roadmap for Semiconductors. Moreover, it is expected that these devices will become more energy efficient in their operation as a result of continued feature size scaling. *(1) Nevertheless, there is a continuing need to dramatically decrease device switching energy over the best scaled projections in order to reduce demands on energy use for computation and sensing.*

One clear implication of the topics addressed in this book is that the nanomorphic cell must perform its functions at the extremes of energy efficiency. Only micro joules are likely to be available from energy sources in the small volume of the nanomorphic cell and this energy must support all cell functions. The examination of potential energy sources given in Chapter 2 emphasized the limits attainable for energy per unit volume and energy per unit mass for a variety of energy sources including the super capacitor, the fuel cell, radio-isotope sources, and various energy-harvesting techniques. Each of these energy sources occupies a different location in the power–energy space. In vivo energy harvesting is an appealing idea since operation of the nanomorphic cell could be extended for a considerable period of time. However, it appears that only a miniscule amount of energy could be captured by most of the known techniques and that this would need to be converted into a form for use by the nanomorphic cell. However, conditioning of the harvested energy would require utilization of some of the very limited volume of the cell. The galvanic cell was chosen as a model energy source in the book but even if all of the 10 micron cube volume is devoted to the galvanic cell, only about one micro joule of energy could be stored. *(2) The need for an adequate supply of energy to support the necessary functions of the nanomorphic cell is a possible show-stopper and creative solutions are needed.*

Another conclusion from the studies in Chapter 5 is that the uniform electronic transmission of data, even for distances on the order of one meter, is very costly from an energy use perspective. This communication challenge is made more difficult by the small size of the cell which dictates that antennas should have dimensions on the order of a few microns. The net result is that directed transmission (not omnidirectional) will probably need to be used at wavelengths on the order of one micron, i.e., the near optical/infrared regime for which device technologies exist for implementation on the scale of nanomorphic cell dimensions. Ideas are needed that enable energy-efficient communication to an external agent for in vivo applications. *(3) Would 'on-the-fly' networks of nanomorphic cells that require largely local communication distances of a few microns between cells offer an energy efficient approach to the communication challenge?*

Chapter 4 on sensors emphasized the importance of selectivity, sensitivity, and registration period in these sign of sensor systems. A first-principles treatment was offered to analyze these parameters for classes of biological sensors including chemical, thermal, and bio-electrical. An encouraging conclusion was that nanowire-based sensors may offer a technology for nanomorphic cell sensor applications that is substantially superior to sensors based on the use of bulk material. *(4) The performance of the nanomorphic cell sensors ultimately determines the usefulness of the cell and research is needed to ensure that both false positives and false negatives are reduced to an absolute minimum.*

The search for existing complex, micron-scale systems with a high degree of functionality and operational energy efficiency against which to benchmark the nanomorphic cell led to consideration of the living cell as a benchmark candidate. The phrases in silico and in carbo were adopted and used to differentiate the underlying material systems used by the nanomorphic and living cell respectively. In order to make the comparison meaningful, a model of the living cell as a computer was developed and compared with the nanomorphic cell. (The term computer used here embraces not only the normal logic and memory elements but also its input/output systems.) The analyses offered in Chapter 6 suggest that the in carbo system is much more energy efficient in the implementation of its computations than the in silico system. This assessment should be viewed as preliminary and the authors encourage further research to sharpen the comparisons offered herein. In any case, it seems clear that innovation is needed to obtain in silico systems that compare favorably with in carbo systems. *(5) This appears to be an area of research where significant breakthroughs in the understanding and therefore the design of in silico (or inorganic systems) could be obtained.*

Almost certainly some form of three-dimensional assembly will be required for the nanomorphic cell. The technology for three-dimensional assembly is just beginning to emerge for integrated circuits, most often taking the form of stacked memory layers attached to the surface of a microprocessor chip and electrically connected to the chip. The conventional assembled system in this case is on the order of centimeters in scale and the assembly accuracies (e.g. placement) are of the order of micrometers. The nanomorphic cell dramatically reduces the size of the assembled system and the problem of testing a fabricated system to which few external connections can be made is indeed challenging. *(6) New assembly (radically different and likely 'bottom-up') technologies are required for the fabrication of the nanomorphic cell if it is to move from 'thought problem' status to physical realization.*

Design tools and system architectures for the nanomorphic cell are likely to be very different from conventional tools and architectures if the desired functionality is to be achieved. For example, the cell is likely to contain functional elements in all three dimensions and the computational architectures that it will employ may be bio-inspired. Moreover, repair of the cell for in vivo applications is not possible, thus forcing an emphasis on the design of reliable systems. *(7) Research is needed to develop three-dimensional computer tools that enable the design and test of the multi-functional nanomorphic cell.*

The focus of the book has been on trade-offs between various information technologies that are needed to support the activities of the nanomorphic cell. However, there are other scientific and technological considerations that might need to be considered, depending on the application. For example, although materials and their properties have been utilized throughout the book, it has not been possible to address materials research in a comprehensive manner. It is clear that advances in materials will further enable micron-scale systems; for example, as indicated herein, it appears that carbon nanotubes could be important for sensor technologies in nanoscale biological applications. In a biological context, the compatibility of materials with living systems remains an important area of research.

In many applications, the nanomechanical capabilities of the cell are important to achieving intended functionality. For example, if the nanomorphic cell requires a means for self-locomotion, then the design space is further complicated by the need to enlarge the energy/volume trade-offs to include nanomechanical systems. It is also conceivable that the nanomorphic cell could be required to take some form of therapeutic action based on its findings, e.g., inject medication, and this could also require the use of nanomechanical actuators. Although not covered in the book, connections to nanomechanics can easily be envisioned for the nanomorphic cell.

This book describes some of the physical considerations and associated limits that are manifest when one considers the design of a system whose dimensions are on the scale of microns. In spite of the many technical challenges, some of which are identified above, the quest for functional micron-size systems does not appear to be beyond consideration. To be sure, the challenges are not easy, but it is hoped that this book will be of value to those who must address the design of micron-scaled systems at the limits of semiconductor technology and that it will inspire creative approaches to these challenges.

Index